An Introduction to Quantum Optics

An open systems approach

IOP Series in Emerging Technologies in Optics and Photonics

Series Editor

R Barry Johnson a Senior Research Professor at Alabama A&M University, has been involved for over 50 years in lens design, optical systems design, electro-optical systems engineering, and photonics. He has been a faculty member at three academic institutions engaged in optics education and research, employed by a number of companies, and provided consulting services.

Dr Johnson is an IOP Fellow, SPIE Fellow and Life Member, OSA Fellow, and was the 1987 President of SPIE. He serves on the editorial board of Infrared Physics & Technology and Advances in Optical Technologies. Dr Johnson has been awarded many patents, has published numerous papers and several books and book chapters, and was awarded the 2012 OSA/SPIE Joseph W Goodman Book Writing Award for Lens Design Fundamentals, Second Edition. He is a perennial co-chair of the annual SPIE Current Developments in Lens Design and Optical Engineering Conference.

Foreword

Until the 1960s, the field of optics was primarily concentrated in the classical areas of photography, cameras, binoculars, telescopes, spectrometers, colorimeters, radiometers, etc. In the late 1960s, optics began to blossom with the advent of new types of infrared detectors, liquid crystal displays (LCD), light emitting diodes (LED), charge coupled devices (CCD), lasers, holography, fiber optics, new optical materials, advances in optical and mechanical fabrication, new optical design programs, and many more technologies. With the development of the LED, LCD, CCD and other electo-optical devices, the term 'photonics' came into vogue in the 1980s to describe the science of using light in development of new technologies and the performance of a myriad of applications. Today, optics and photonics are truly pervasive throughout society and new technologies are continuing to emerge. The objective of this series is to provide students, researchers, and those who enjoy self-teaching with a wide-ranging collection of books that each focus on a relevant topic in technologies and application of optics and photonics. These books will provide knowledge to prepare the reader to be better able to participate in these exciting areas now and in the future. The title of this series is Emerging Technologies in Optics and Photonics where 'emerging' is taken to mean 'coming into existence,' 'coming into maturity,' and 'coming into prominence.' IOP Publishing and I hope that you find this Series of significant value to you and your career.

An Introduction to Quantum Optics

An open systems approach

Perry Rice
Autonomy, Artificial Intelligence, and Machine Learning Lab, Azimuth Corporation, 2970 Presidential Dr., Suite 200 Fairborn, OH 45324, USA

IOP Publishing, Bristol, UK

ISBN 978-0-7503-1713-9 (ebook)
ISBN 978-0-7503-1711-5 (print)
ISBN 978-0-7503-1809-9 (myPrint)
ISBN 978-0-7503-1712-2 (mobi)

DOI 10.1088/978-0-7503-1713-9

Version: 20200901

IOP ebooks

British Library Cataloguing-in-Publication Data: A catalogue record for this book is available from the British Library.

Published by IOP Publishing, wholly owned by The Institute of Physics, London

IOP Publishing, Temple Circus, Temple Way, Bristol, BS1 6HG, UK

US Office: IOP Publishing, Inc., 190 North Independence Mall West, Suite 601, Philadelphia, PA 19106, USA

This book is dedicated to the memory of my dad, Paul Richard Rice.
A kind and loving man with a great fondness for music.

Contents

Preface

In this volume, we hope to introduce some of the mathematical fundamentals of quantum optics, from a statistical physics point of view. The material is meant to be understandable and usable by upperclassmen and beginning graduate students. Chapter 1 is a brief review of what is meant, what I mean, by quantum optics. The next two chapters, chapters 2 and 3, are reviews of classical electromagnetism and quantum mechanics. The reviews are not comprehensive, but provide the tools we shall need to proceed. Next in chapter 4, we turn to the interaction of a simple two-level system with an electromagnetic wave to begin our studies of emission and absorption. The basics of the quantum treatment of the electromagnetic field are presented in chapter 5. In chapter 6, we revisit the simple two-level system, but let it interact with a quantized electromagnetic field. Measurement plays a key role in quantum optics, and in quantum mechanics in general. Various detection schemes are discussed in chapter 7. The heart of the matter is in chapters 8 and 9. Chapter 8 introduces the density matrix formulation of quantum mechanics for an open quantum system; one that exchanges energy and information with its environment. Chapter 9 discusses quantum trajectory theory, a manner of unraveling the density matrix to allow for ease of computation as well as intuition. This is particularly useful for undergraduates, who understand evolution of quantum wave functions and do not shy away from the need for quantum jumps. Finally, we conclude in chapter 10 with the use of quasiprobability functions to describe quantum optical systems.

This book is the result of teaching quantum optics to undergraduates and master's students for 30 years at Miami University. The purpose was to introduce students to the mathematical formalism of quantum optics in the most physical way I knew how.

Perry Rice, Xenia, March 14, 2020

Acknowledgments

There are far too many students to thank for suffering through the development of this book over that time but I must thank my MS advisees Xiaosong Yin, Tim Burt, Bobby Jones, Pranaw Rungta, James Clemens, Puneet Swarup, Elliot Strimbu, Joe Leach, Mambwe Mamba, Dyan Jones, Jeff Hyde, Habtom Woldekristos, Todd vanWoerkom, Charlie Baldwin, Thomas Jenkins, Jared Goettemoeller, Robert McCutcheon, Tyler Thurtell, Arkan Hassan, Daniel King, and Mitch Mazzei. Also, Matt Terraciano and Joe Reiner from other institutions. Undergraduate research students that were helpful in the development of this book were James Walden, Chris Baird, Shohini Ghose, Scott Secrest, Jeffrey Hyde, Luke Keltner, Bill Konyk, Robert Rogers, and Jason Rook. Most of these students produced research articles using the techniques described within.

I must also thank my undergraduate mentors at Wright State University, Merrill Andrews, Gust Bambikidis, Joe Hemsky, Mitchell Simpson, and David Wood. The attention given to their students was the beginning of my development as a teacher. In graduate school at the University of Arkansas, I had the pleasure of interacting with Howard Carmichael, James Cresser, William Harter, Peter Milonni, Surendra Singh, and Reeta Vyas. One could not have asked for a better set of teachers and role models. Fellow travellers of note were Paul Alsing, Greg Holliday, Steve Montgomery, John Shultz, Jan Yarrison, David Weeks, and Murray Wolinsky. I have been fortunate to have met Luis Orozco at this time and begin a lifetime of collaboration and friendship.

My interest in quantum optics began when my colleagues Leno Pedrotti went to graduate school working with Marlon Scully and Robert Brecha worked with H Jeffrey Kimble. Leno and Robert have remained good friends and quantum colleagues over the years and I have treasured that. I was working for the USAF wondering what to do when I discovered Richard Cook, who patiently worked through Glauber's initial papers, and directed me to the University of Arkansas and Peter Milonni. Before Peter moved to Los Alamos, I was able to learn much about the nature of the quantum vacuum, and how beautiful quantum electrodynamics is, and quantum field theory in general. My research advisor was Howard Carmichael. He perhaps has had the most impact on how I view the quantum world. One is biased of course, but I think he is the finest advisor and friend one can have.

Finally, my loving family who started me out needs notice. My dad Paul, mom Lucy, grandma Gertrude, aunt Fern, and Uncle Bud. At present my wife Jan and son Patrick are at my side. I feel very lucky to have walked the path I have, with the company I had.

And as they say, all mistakes in the book are solely attributable to me. I hope this is of value to people starting to learn about the mysteries of quantum optics.

Author biography

Perry Rice

Dr Perry Rice earned a BS in Physics at Wright State University in 1981 and a PhD in Physics from the University of Arkansas in 1988. He then spent 30 years as a professor at Miami University, followed by a year at the Oregon Center for Optical, Molecular, and Atomic Physics. His research is in the interaction of quantized light with matter, and properties of quantum noise, including resonance fluorescence, cavity quantum electrodynamics, and nonlinear optics. He is currently a senior research scientist at Azimuth Corporation using neural networks to work on waveguide quantum electrodynamics, error correction codes, and distribution of multipartite entanglement. In his spare time, he enjoys playing and listening to jazz, blues, folk, and rock music.

IOP Publishing

An Introduction to Quantum Optics
An open systems approach
Perry Rice

Chapter 1

Introduction

1.1 What is quantum optics

The simplest 'definition' of quantum optics is the discipline that examines 'what is a photon'. A more detailed description would be a 'non-relativistic quantum field theory of atoms interacting with light'. The field grew out of the invention of the laser and the question of 'what is the difference between laser light and a light bulb'. The acronym LASER stands for 'light amplification by stimulated emission of radiation'. Often in textbooks it is stated that laser light is *directional, intense, and monochromatic.* Well I can make light directional via a fiber optic, I can turn up the power until the thing melts, and I can filter the output to limit the bandwidth. But what about stimulated emission? When a photon interacts with a detector, it does not announce its origin as stimulated or spontaneous, it is just an excitation of an electromagnetic field mode. My definition of a photon is a click in an absorptive detector, as we will detail later. It is not a particle, or a wave, but an excitation of a bosonic field. So, how DO you tell whether the light is coming from an incandescent bulb or a HeNe laser? Glauber [9] taught us that the difference is in *photon correlations.* One must look past expectation values and examine two-time correlations at the least. The fluctuations in the light field provide information that cannot be accessed by just examining averages. It will turn out that for laser light the statistics of the photo detection events are Poissonian, each one independent of the other. For light from an incandescent bulb, thermal light, we find that the photons are more likely to come in pairs than they would be for a random source. This is the origin of the Hanbury–Brown–Twiss effect [10], whereby one can detect correlations in intensity from two spatially separated detectors. We will find that light sources exist where the photons are less likely to come in pairs than from a laser, which is called photon anti-bunching. This type of light is quite interesting as it *cannot* be modeled by a deterministic random process. A cartoon of this concept is shown in figure 1.1.

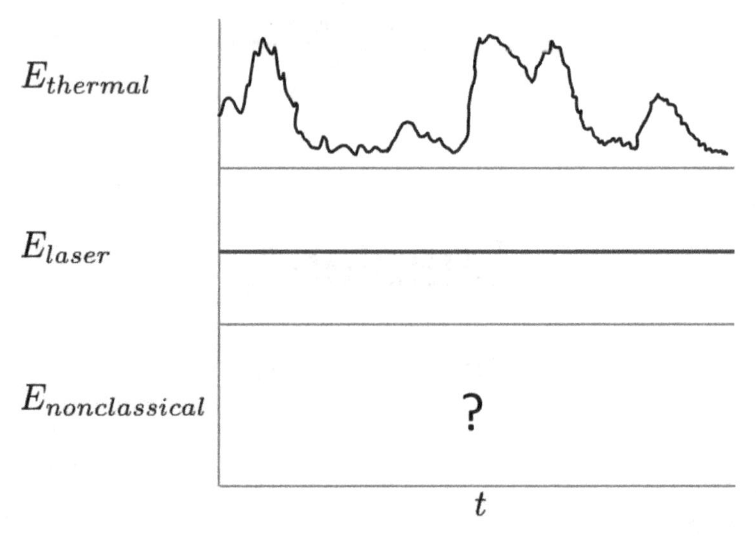

Figure 1.1. Electric field amplitude time evolution for (a) noisy, thermal source, (b) laser source, and (c) nonclassical or manifestly quantum light source.

In thermal light, there are large fluctuations, and during periods of higher intensity you are more likely to detect a photon. You are also more likely to get a second photon, simply because then there are more photons. For a laser, the output is described by a Poisson stochastic process, after detection of a photon you are no more or less likely to detect a second photon. A fully quantum theory of the laser due to Scully and Lamb [20] showed the Poissonian nature of the photon statistics and pointed out the key role of spontaneous emission in determining the non-zero linewidth of the laser output. What about anti-correlations? The simplest example of such a source is a single two-level atomic transition driven by a laser, and examining the fluorescence. To spontaneously emit, the atom must be excited. Upon emission the atom goes to the ground state and hence cannot emit a second photon. The photon statistics and spectra of this system of resonance fluorescence was the focus of intense work in the mid-1970s, and a fully quantum model of the electromagnetic field was found to be necessary to understand and predict results of experiments.

How can we differentiate different field states if we cannot measure the field directly? Let us consider a field inside a box and ask how many photons are in the box. We will open the box and let all the photons leak out and count how many we detect. Of course, we assume perfect photo detection for now. Then let us do that over and over, as our answer will be a random number generally. Different field states will have different statistics and different probability distributions for the number of photons we detect. Examples of such distributions are shown in figure 1.2. There are states of the field with a definite number of photons, called number, or Fock states. The coherent state that is output from a laser has a Poissonian distribution, and a thermal field has the decaying exponential distribution

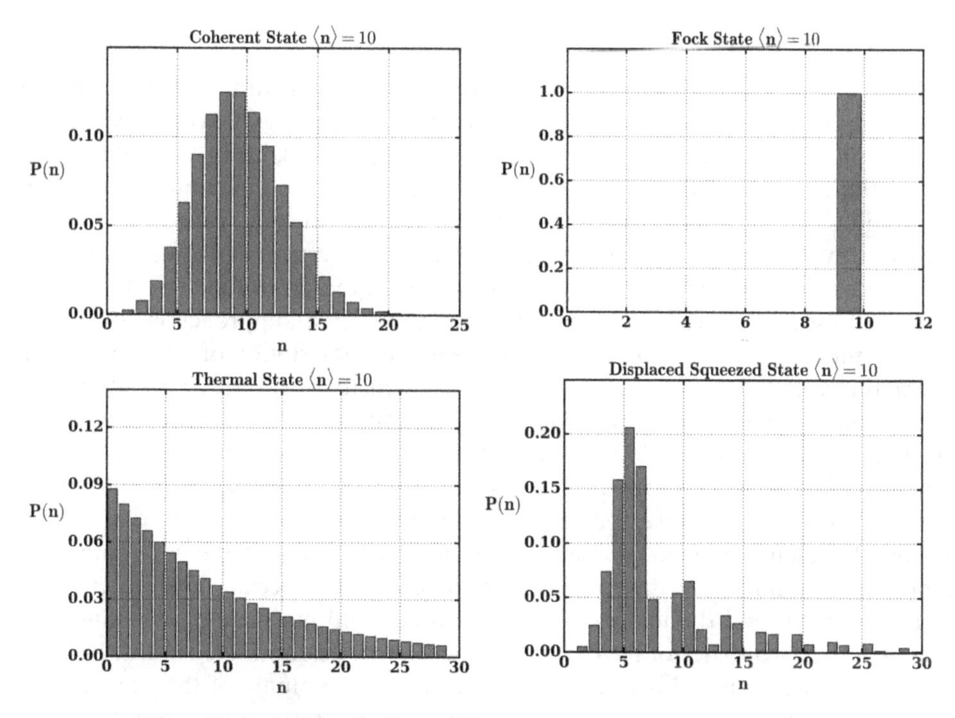

Figure 1.2. Probability distribution of photons $P(n)$ for the given field states.

characteristic of a Bose–Einstein distribution. More interesting distributions result from nonclassical states such as a displaced squeezed vacuum or squeezed state. How do we prepare such states in a box? Good question. What we are generally more interested in is a slightly leaky box, where we also inject photons. The result is often a steady state that is not in thermal equilibrium, a non-equilibrium steady state. The leaky box can be two mirrors forming an optical cavity, such as in a Fabry–Perot or a semiconductor wave guide with polished ends. We can use an essentially classical source, the output of a laser. We can put single atoms, a gas of many atoms, or crystals inside this slightly leaky box and create field states of our choice. Some are more difficult than others of course. We can infer what is inside the box by the statistical properties of what leaks out. This could be through a non-perfect mirror, or fluorescence out the sides of the cavity, perhaps via spontaneous emission. The leaky part is important, as we would like to create a field source, and the leak is essentially our output. A laser with two perfect mirrors would generate a large field inside the cavity, but if no light comes out, we cannot make laser pointers, blow holes in things, or do laser spectroscopy.

The focus of this book is to examine how to calculate the statistics of photo counts, as well as various spectral quantities in a variety of systems. We will mainly look at two-time correlations, but of course the probability distribution is determined in full by the entire set of n-time correlation functions, just as in quantum electrodynamics and other field theories.

1.2 Open quantum systems

In this example above, resonance fluorescence, the atom is not an isolated system. Energy is put into the system via the driving laser, and energy leaves the system via interaction with the field modes (all directions, frequencies, and polarizations) that are in the ground or vacuum state. It is via the latter interaction that quantum fluctuations, or vacuum fluctuations, enter the picture. Such a system that interacts with its environment is known as an *open quantum system*. Such a system is not described by a wave function, unless one considers the wave function of the atom and all of the field modes, an infinite number. So, such an approach is not tenable, and we must resort to approximations. If we want a description of our atom and its interaction with its environment, we must use a density operator, or density matrix. In the simplest models, the density operator evolves via the Schrödinger supplemented by corrections due to the environmental interactions. These corrections will depend only on gross properties of the environment, such as temperature and pressure. In this procedure, we reduce the number of equations by omitting the details of the environment, thereby losing information about the environment. A good example might be dropping an ice cube into the ocean. The ice cube will eventually melt, and the temperature of the ocean will not change very much, nor will any other properties of the ocean. The ocean can be considered a heat bath at a constant temperature. This is mainly due to the large volume of the ocean, an ice cube dropped into a glass of water will lower the temperature of the resulting water by a measurable amount, and as such thermodynamic properties of water can be measured via calorimetry. In the case of the quantum electromagnetic vacuum, we have an infinite number of modes. An infinite number of frequencies, each with an infinite number of propagation directions and two polarizations. Therefore, it seems reasonable that an approach based on an interaction of a closed quantum system with a type of heat bath will be useful.

We consider the coupling of the atom to a given field mode (g) as very small, that it scales with the inverse square root of the volume of the field mode (which is infinite). But the number of modes N is infinite. We will find that the spontaneous emission rate of the atom depends on the product Ng^2 which can take on a finite value, as in figure 1.3. The fancy title for what we are doing is quantum statistical mechanics.

1.3 This book

This book resulted from notes developed for teaching quantum optics to upper level undergraduates, and grad students starting out in the field, or anyone with a good grasp of quantum mechanics that wishes to know more. Quantum optics is one of the seminal fields for the revolution underway in quantum information processing, detection of gravitational waves, many-body quantum systems, nanophotonics, and others. As such, an audience beyond specialists will want/need to understand the basics of quantum optics. This book seeks to present the simplest non-trivial models that illustrate the principles and is not meant to be exhaustive. In the 1980s while in graduate school, the key books (at least for me) were those of Loudon [11]; Allen

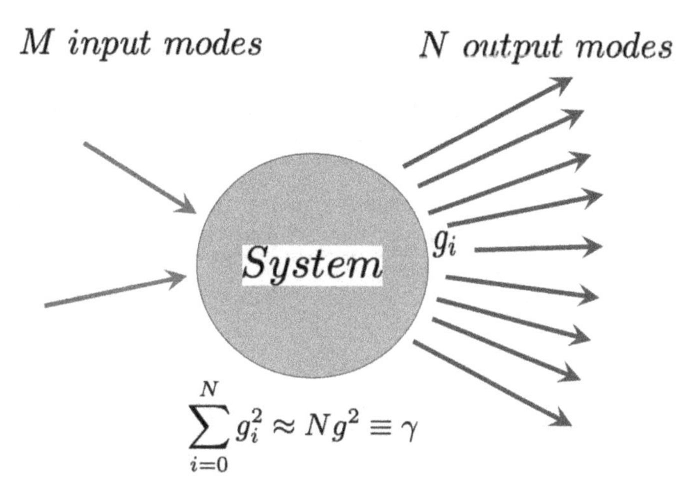

Figure 1.3. Sketch of an open quantum system with a (typically small) M number of input modes and a large number N of output modes. The coupling of the system to a given mode is g_i.

and Eberly [2]; and Sargent, Scully, and Lamb [19]. I would suggest that the modern trifecta would include the books by Orszag [18], Scully and Zubairy [21], and Mandel and Wolf [12]. I would also recommend the trilogy written by Carmichael [4–6] and the work of Milonni [14–17] for more details. My interactions with the latter two largely form the basis of my approach. A similar approach is that of Bruer and Pettrucione [3]. Other books I am familiar with enough to highly recommend are those of Walls and Milburn [23], Gardiner and Zoller [7], Gerry and Knight [8], Meystre and Sargent [13], and Agarwal [1]. I would be remiss not to direct the readers' attention to the set of notes on quantum and atom optics notes available online, provided by Steck [22]. There are many good texts on quantum optics these days, so why one more? The novelty hopefully is the accessibility of the material to a newcomer to the field, and it is more pedagogical than exhaustive.

References

[1] Agarwal G 2012 *Quantum Optics* (Cambridge: Cambridge University Press)
[2] Allen L and Eberly J 1975 *Optical Resonance and Two-Level Atoms* (New York: Wiley)
[3] Breuer H-P and Petruccione F 2002 *The Theory of Open Quantum Systems* (Oxford: Oxford University Press)
[4] Carmichael H J 1993 *An Open Systems Approach to Quantum Optics* (Berlin: Springer)
[5] Carmichael H J 2007 *Statistical Methods in Quantum Optics. 2: Non-Classical Fields* Theoretical and Mathematical Physics (Berlin: Springer)
[6] Carmichael H J 1999 *Statistical Methods in Quantum Optics. 1: Master equations and Fokker–Planck Equations* Physics and Astronomy Online Library (Berlin: Springer)
[7] Gardiner C and Zoller P 2004 *Quantum Noise* (Berlin: Springer)
[8] Gerry C and Knight P 2004 *Introductory Quantum Optics* (Cambridge: Cambridge University Press)
[9] Glauber R 1963 Coherent and incoherent states of the radiation field *Phys. Rev.* **131** 2766–88

[10] Hanbury-Brown R and Rwiss R Q 1956 Correlation between photons in two coherent beams of light *Nature* **177** 27–9
[11] Loudon R 2000 *The Quantum Theory of Light* (Oxford: Oxford University Press)
[12] Mandel L and Wolf E 1995 *Optical Coherence and Quantum Optics* (EBL-Schweitzer: Cambridge University Press)
[13] Meystre P and Sargent M 2007 *Elements of Quantum Optics* (Berlin, Heidelberg: Springer)
[14] Milonni P 2004 *Fast Light, Slow Light, and Left-handed Light* Optics and Optoelectronics (Beograd: Institute of Physics)
[15] Milonni P 2019 *An Introduction to Quantum Optics and Quantum Fluctuations* (Oxford: Oxford University Press)
[16] Milonni P W 1994 *The Quantum Vacuum: An Introduction to Quantum Electrodynamics* (New York: Academic)
[17] Milonni P W and Eberly J H 2010 *Laser Physics* (New York: Wiley)
[18] Orszag M 2000 *Quantum Optics* Advanced Texts in Physics (Berlin: Springer)
[19] Sargent M, Scully M and Lamb W 1975 *Laser Physics* (Reading, MA: Addison-Wesley)
[20] Scully M O and Lamb W E 1967 Quantum theory of an optical maser. I. General theory *Phys. Rev.* **159** 208–26
[21] Scully M O and Zubairy M S 1997 *Quantum Optics* (Cambridge: Cambridge University Press)
[22] Steck D Atom and quantum optics notes (course notes) http://atomoptics-nas.uoregon.edu/~dsteck/teaching/quantum-optics/
[23] Walls D and Milburn G 2008 *Quantum Optics* (Berlin: Springer)

IOP Publishing

An Introduction to Quantum Optics

An open systems approach

Perry Rice

Chapter 2

Classical electromagnetism and linear optics

2.1 Maxwell equations and electromagnetic waves

Here, we review the basics of electromagnetism and applications to wave propagation. Standard references would include [1–6]. We begin with a discussion of Maxwell's equations. Recall that these are

Gauss' law	$\vec{\nabla} \cdot \vec{E} = \rho/\varepsilon_0$
The law with no name	$\vec{\nabla} \cdot \vec{B} = 0$
Faraday's law	$\vec{\nabla} \times \vec{E} = \frac{-\partial \vec{B}}{\partial t}$
Ampère's law	$\vec{\nabla} \times \vec{B} = \mu_0 \vec{J} + \frac{1}{c^2}\frac{\partial \vec{E}}{\partial t}$

In these, we have the total charge density ρ and the total current density $\vec{J}$, as sources for the electric and magnetic fields, $\vec{E}$ and $\vec{B}$. Also, we have defined $c^2 = 1/\varepsilon_0\mu_0$. The second equation, the one *not* named after a dead white European male, is automatically satisfied if we have

$$\vec{B} = \vec{\nabla} \times \vec{A} \tag{2.1}$$

as we have the identity that $\vec{\nabla} \cdot (\vec{\nabla} \times \overrightarrow{anything}) = 0$; the divergence has no curl. This allows us to introduce $\vec{A}$, the vector potential. The usual electric potential emerges by using equation (2.1) in Faraday's law,

$$\begin{aligned}\vec{\nabla} \times \vec{E} &= \frac{-\partial \vec{B}}{\partial t} \\ &= -\vec{\nabla} \times \frac{\partial \vec{A}}{\partial t}\end{aligned} \tag{2.2}$$

which allows us to write

$$\vec{E} = -\frac{\partial \vec{A}}{\partial t} + \vec{\eta} \tag{2.3}$$

where we must have $\vec{\nabla} \times \vec{\eta} = 0$. We can now make use of the identity $\vec{\nabla} \times \vec{\nabla}(\textit{anyscalar function}) = 0$, or the gradient has no divergence, to write

$$\vec{\eta} = -\vec{\nabla}\phi \tag{2.4}$$

where ϕ is the usual electric potential. We now have a relation between the potentials and the fields

$$\vec{E} = -\frac{\partial \vec{A}}{\partial t} - \vec{\nabla}\phi \tag{2.5}$$

$$\vec{B} = \vec{\nabla} \times \vec{A} \tag{2.6}$$

Let us note that we now have four variables (functions really), the three components of the vector potential and the scalar potential. How many equations do we have? Well TLWNN and Faraday's law are automatically satisfied by the potentials, they can be viewed as constraints akin to initial or boundary conditions. Thus, we have four first-order equations (one vector equation in Ampère's law and one scalar equation, Gauss' law) and four constraints, which allows us to have a solution. The potentials are *not* unique as we shall see. Consider the following transformation, known as a gauge transformation

$$\vec{A}' = \vec{A} - \vec{\nabla}\lambda \tag{2.7}$$

$$\phi' = \phi + \frac{\partial \lambda}{\partial t} \tag{2.8}$$

What fields do these potentials produce?

$$\begin{aligned} \vec{B}' &= \vec{\nabla} \times \vec{A}' \\ &= \vec{\nabla} \times \vec{A} \\ &= \vec{B} \end{aligned} \tag{2.9}$$

$$\begin{aligned} \vec{E}' &= -\vec{\nabla}\phi' - \frac{\partial \vec{A}'}{\partial t} \\ &= -\vec{\nabla}\phi - \frac{\partial}{\partial t}\vec{\nabla}\lambda - \frac{\partial \vec{A}}{\partial t} + \frac{\partial}{\partial t}\vec{\nabla}\lambda \\ &= -\frac{\partial \vec{A}}{\partial t} - \vec{\nabla}\phi \\ &= \vec{E} \end{aligned} \tag{2.10}$$

So, we see that our electric and magnetic fields will be unchanged by the gauge transformation, thus it would seem that the potentials are not unique. What about the fields themselves? They are unique, a fact guaranteed by Helmholtz's theorem which states that the curl and divergence of a vector field uniquely defines that field.

So, we can use the potentials instead of the fields; in introductory physics, the electric potential is introduced early on as it is often easier to solve for a scalar function and then takes its gradient to obtain the electric field. There is no such utility with the vector potential, hence it appears later! But what equations do these potentials satisfy? Let us start down a well trodden path by examining Faraday's equation in terms of potentials, We now need to find out more about the vector and scalar potentials. We do this by examining what equations they need to satisfy. We have

$$\vec{\nabla} \times \vec{B} = \vec{\nabla} \times (\vec{\nabla} \times \vec{A}) = \frac{1}{c^2}\left(\vec{\nabla}\frac{\partial \phi}{\partial t} - \frac{\partial^2 \vec{A}}{\partial t^2}\right) + \mu_0 \vec{J} \tag{2.11}$$

and

$$\vec{\nabla}(\vec{\nabla} \cdot \vec{A}) - \nabla^2 \vec{A} + \frac{1}{c^2}\vec{\nabla}\frac{\partial \phi}{\partial t} + \frac{1}{c^2}\frac{\partial^2 \vec{A}}{\partial t^2} = \mu_0 \vec{J} \tag{2.12}$$

$$\vec{\nabla} \cdot \vec{E} = -\vec{\nabla} \cdot (\vec{\nabla}\phi) - \frac{\partial}{\partial t}(\vec{\nabla} \cdot \vec{A}) = \frac{\rho}{\varepsilon_0} \tag{2.13}$$

So that when we combine all of these, the two resultant equations are given as

$$\vec{\nabla}(\vec{\nabla} \cdot \vec{A}) - \nabla^2 \vec{A} + \frac{1}{c^2}\vec{\nabla}\frac{\partial \phi}{\partial t} + \frac{1}{c^2}\frac{\partial^2 \vec{A}}{\partial t^2} = \mu_0 \vec{J} \tag{2.14}$$

$$-\nabla^2 \phi - \frac{\partial}{\partial t}(\vec{\nabla} \cdot \vec{A}) = \frac{\rho}{\varepsilon_0} \tag{2.15}$$

Now, we want to work in the Coulomb gauge, $\vec{\nabla} \cdot \vec{A} = 0$. This can be explicitly shown by letting $\vec{A} \longrightarrow \vec{A}_0 - \vec{\nabla}\eta$ and $\phi \longrightarrow \phi_0 + \partial\eta/\partial t$ for any function η. The Coulomb gauge is defined by $\vec{\nabla} \cdot \vec{A} = 0$. This gauge makes our equations a bit simpler, because since

$$\nabla^2 \eta = \vec{\nabla} \cdot \vec{A} \tag{2.16}$$

we can write our two equations for the potentials as

$$\nabla^2 \vec{A} - \frac{1}{c^2}\frac{\partial}{\partial t}\vec{\nabla}\phi = -\mu_0 \vec{J} \tag{2.17}$$

$$\nabla^2 \phi = -\frac{\rho}{\varepsilon_0} \tag{2.18}$$

The scalar potential satisfies Poisson's equation with static solution

$$\phi(\vec{r}) = \frac{1}{4\pi\varepsilon_0}\int \frac{\rho(\vec{r}\,')}{|\vec{r} - \vec{r}\,'|}d^3\vec{r}\,' \tag{2.19}$$

We next decompose the current $\vec{J}$ into transverse and longitudinal terms $\vec{J} = \vec{J}_T + \vec{J}_L$ so that

$$\vec{\nabla} \times \vec{J}_L = 0$$
$$\vec{\nabla} \cdot \vec{J}_T = 0$$

Now, since the curl of $\vec{J}_L = 0$, we can write $\vec{J}_L = \vec{\nabla}\psi$

We then have

$$\vec{\nabla} \cdot \vec{J}_L = -\frac{\partial \rho}{\partial t} \tag{2.20}$$

and

$$\vec{\nabla} \cdot (\vec{\nabla} \times \vec{B}) = \varepsilon_0 \vec{\nabla} \cdot \frac{\partial \vec{E}}{\partial t} + \vec{\nabla} \cdot \vec{J} \tag{2.21}$$

The longitudinal current density satisfies

$$\vec{\nabla} \cdot \vec{J}_L = \vec{\nabla} \cdot (\vec{\nabla}\psi) \tag{2.22}$$

$$= \nabla^2\psi \tag{2.23}$$

$$= -\frac{\partial \rho}{\partial t} \tag{2.24}$$

This is nonzero if the charges are moving transverse to the direction of propagation for some reason. We then have

$$\psi = \varepsilon_0 \frac{\partial \phi}{\partial t} \tag{2.25}$$

$$\vec{J}_L = \varepsilon_0 \vec{\nabla} \frac{\partial \phi}{\partial t} \tag{2.26}$$

Finally then, we have

$$-\nabla^2 \vec{A} + \frac{1}{c^2}\frac{\partial^2 \vec{A}}{\partial t} = \mu_0 \vec{J}_T \tag{2.27}$$

with the form of the solution being

$$\vec{A} = \frac{\mu_0}{4\pi}\int \frac{\vec{J}_T(\vec{r}\,', t)}{|\vec{r} - \vec{r}\,'|}d^3\vec{r}\,' \tag{2.28}$$

The equations that determine the scalar potential are

$$\nabla^2 \phi = \frac{\rho}{\varepsilon_0} \tag{2.29}$$

$$\varepsilon_0 \vec{\nabla} \frac{\partial \phi}{\partial t} = \vec{J}_L \tag{2.30}$$

2.2 Wave equation for fields in a medium

Let us now consider Maxwell's equations in a medium. These would be

Gauss' law	$\vec{\nabla} \cdot \vec{D} = \rho_{\text{free}}$
The law with no name	$\vec{\nabla} \cdot \vec{B} = 0$
Faraday's law	$\vec{\nabla} \times \vec{E} = \frac{-\partial \vec{B}}{\partial t}$
Ampère's law	$\vec{\nabla} \times \vec{H} = \mu_0 \vec{J}_{\text{free}} + \frac{\partial \vec{D}}{\partial t}$

where we also have the constitutive relations

$$\vec{D} = \varepsilon_0 \vec{E} + \vec{P} \tag{2.31}$$

$$\vec{B} = \mu_0 \vec{H} + \vec{M} \tag{2.32}$$

with the polarization of the medium $\vec{P}$ and magnetization $\vec{M}$, and the sources are the free charges and free currents. These relations are heuristic in nature and not fundamental. In this book, we will consider nonmagnetic materials and set $\vec{M} = 0$. To see how a polarization may arise, consider an electrically neutral atom or molecule. If an external field is applied, the atomic nucleus will want to move along the direction of the field as it is positively charged, and the electrons will want to move in the opposite direction. This leads to a separation of charge, with a resultant field that opposes the applied field. This is a dipole field, with $\vec{P} = N\vec{\mu} = Ne\vec{x}$ for the case of N atoms, with $\vec{x}$ being the separation between the nucleus and the electron, for an alkali element, for example. In a linear, isotropic, homogeneous material, the polarization is given by $\vec{P} = \varepsilon_0 \chi \vec{E}$. The quantity χ is the susceptibility of the material. Generally, the situation is more complicated than that. Consider a material with a different response along different directions, an anisotropic material.

If we apply a field E along the x axis, as shown in figure 2.1, we find a polarization $P_x = \varepsilon_0 \chi_x E$. If instead, we apply the field along the y axis, we find a polarization $P_y = \varepsilon_0 \chi_y E$. In both cases, the polarization is in the same direction as the applied field. But what if we apply a field E aligned at a 45 degree angle? What is the resulting polarization? We can find it simply by vector addition, but the result is that the polarization is no longer along the direction of the applied field. To handle such situations, we must promote the susceptibility χ from a scalar to a matrix (technically

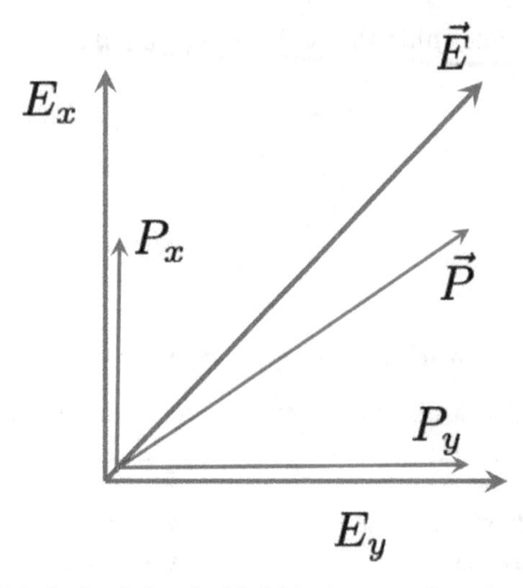

Figure 2.1. Polarization induced with fields along x and y axis, and resultant.

a tensor, more on that later). We can then write the relationship between the polarization and the applied electric field

$$\vec{P} = \varepsilon_0 \chi \vec{E} \tag{2.33}$$

Here, we see that χ transforms or maps the vector $\varepsilon_0 \vec{E}$ into a new vector $\vec{P}$, and the two vectors need not be collinear. In the case of a homogeneous material (i.e. no directional dependence), we have $\chi = \chi \mathbf{I}$, where the scalar χ is the usual susceptibility and $\mathbf{I}$ is the identity matrix. In this case, $\vec{E}$ and $\vec{P}$ are collinear. In the case illustrated in figure 2.1,

$$\chi = \begin{pmatrix} \chi_x & 0 \\ 0 & \chi_y \end{pmatrix} \tag{2.34}$$

and we have in component form

$$P_x = \varepsilon_0 \chi_x E_x \tag{2.35}$$

$$P_y = \varepsilon_0 \chi_y E_y \tag{2.36}$$

More generally, the susceptibility matrix (soon to be promoted to tensor status) has non-zero off-diagonal matrix elements

$$\chi = \begin{pmatrix} \chi_{xx} & \chi_{xy} \\ \chi_{yx} & \chi_{yy} \end{pmatrix} \tag{2.37}$$

or in component form

$$P_x = \varepsilon_0 (\chi_{xx} E_x + \chi_{xy} E_y) \tag{2.38}$$

$$P_y = \varepsilon_0(\chi_{yx}E_y + \chi_{yy}E_y) \tag{2.39}$$

There are special directions in space where the polarization *will* be collinear with the applied field. If the field is applied along one of these special directions, we can replace the susceptibility matrix with a scalar. These special directions will be determined by the eigenvectors of χ, and the value of the susceptibility for that direction will be the corresponding eigenvalue. The situation may be even more complicated; if the material is non-homogeneous, the susceptibility matrix elements may depend on position. Furthermore, for a non-linear material, the susceptibility matrix elements may be dependent on field strength

$$\chi_{ij} \equiv \chi_{ij}(\vec{r}, \vec{E}) \tag{2.40}$$

In fact, the simplest quantum system, a two-level atom, will have a nonlinear susceptibility. All materials will respond nonlinearly at some applied field level, if nothing else an atom will ionize at some point. It is oftentimes useful to express the polarization as a Taylor series in applied field strength

$$P(E) = \varepsilon_0(\chi^{(0)} + \chi^{(1)}E + \chi^{(2)}E^2 + \chi^{(3)}E^3 + \cdots) \tag{2.41}$$

Here, we consider a component of P and E for an isotropic, but nonlinear, material. $\chi^{(0)}$ is related to a permanent dipole moment, $\chi^{(1)}$ is the usual linear susceptibility, and terms like $\chi^{(2)}$ and $\chi^{(3)}$ represent nonlinear terms. If we have a material with inversion symmetry, reversing the direction of the applied field should also reverse the direction of the polarization. We would then have

$$\begin{aligned} P(-E) &= \varepsilon_0(\chi^{(0)} + \chi^{(1)}(-E) + \chi^{(2)}(-E)^2 + \chi^{(3)}(-E)^3 + \cdots) \\ &= \varepsilon_0(\chi^{(0)} - \chi^{(1)}E + \chi^{(2)}E^2 - \chi^{(3)}E^3 + \cdots) \\ &= -P(E) \end{aligned} \tag{2.42}$$

For this to hold true, we must have $\chi^{(0)} = \chi^{(2)} = 0$ as well as all even terms. So, for an isotropic material, we cannot have a permanent dipole moment, and the nonlinearity is cubic at best.

Let us now consider the wave equation for electric fields in a material. We start by taking the curl of Faraday's law

$$\begin{aligned} \vec{\nabla} \times (\vec{\nabla} \times \vec{D}) &= -\vec{\nabla} \times \frac{\partial \vec{B}}{\partial t} \\ &= -\nabla^2 \vec{D} + \vec{\nabla}(\vec{\nabla} \cdot \vec{D}) \\ &= -\mu_0 \vec{J} + \mu_0 \frac{\partial \vec{D}}{\partial t} \end{aligned} \tag{2.43}$$

Utilizing the constitutive relation $\vec{D} = \varepsilon_0 \vec{E} + \vec{P}$, we find

$$\nabla^2 \vec{E} - \frac{1}{c^2}\frac{\partial^2 \vec{E}}{\partial t^2} = \mu_0 \frac{\partial^2 \vec{P}}{\partial t^2} + \mu_0 \frac{\partial \vec{J}}{\partial t} \tag{2.44}$$

where we have taken $\nabla \cdot \vec{P} = 0$. In free space, we have

$$\nabla^2 \vec{E} - \frac{1}{c^2}\frac{\partial^2 \vec{E}}{\partial t^2} = 0 \tag{2.45}$$

and any vector with components of the form $f(k(x - ct)) = f(kx - \omega t)$ is a solution. This is a wave that propagates at speed $c = \sqrt{1/\varepsilon_0 \mu_0}$, with wavelength $\lambda = 2\pi/k$ and angular frequency $\omega = ck$. The source terms on the right-hand side generate electromagnetic waves. As $\vec{P} = Ne\vec{x}$ and $\vec{J} \sim \vec{v}$, both of these terms are proportional to the acceleration of a charge, hence as you have been long told, accelerating charges radiate electromagnetic radiation. Sometimes, we wish to take into account some loss mechanism for the field, and this can be accomplished mathematically by taking the current density to be that of an Ohmic material, $\vec{J} = \sigma \vec{E}$, where σ is the electrical conductivity. This leads to a wave equation of the form

$$\nabla^2 \vec{E} - \frac{1}{c^2}\frac{\partial^2 \vec{E}}{\partial t^2} - \mu_0 \sigma \frac{\partial \vec{E}}{\partial t} = \mu_0 \frac{\partial^2 \vec{P}}{\partial t^2} \tag{2.46}$$

In the case of an electromagnetic field confined to a cavity, this Ohmic loss will in reality stem from light leaking out of an imperfect mirror. Let us ignore losses at this point and consider a linear material with $\vec{P} = \varepsilon_0 \chi \vec{E}$, where χ is the linear susceptibility. We then have

$$\nabla^2 \vec{E} - \frac{1 + \chi}{c^2}\frac{\partial^2 \vec{E}}{\partial t^2} = 0 \tag{2.47}$$

This is just like the free space case, only now the speed of the wave is $v = c/n$ where we have defined an index of refraction $n = \sqrt{1 + \chi}$. Waves would be of the form $E = \exp(i(Kx - \omega t))$, where $K = nk$. We consider the medium to be linear, and so it responds at the frequency it is driven at. Since $v = c/n = f\lambda$, if f does not change, then the wavelength must decrease, $\lambda \to \lambda/n$, or $k \to nk$. Let us consider what would happen if the index of refraction were complex, $n = n_R + in_I$. In this case, wave solutions would take the form

$$E = \exp(i(n_R kx - \omega t)) \exp(-n_I kx) \tag{2.48}$$

where k is the free space wavenumber. We see that the real part of the index of refraction is what we are used to having, something that slows the wave down and shortens its wavelength. If there is an imaginary part, we have loss as the wave propagates if $n_I > 0$, this is a way to model absorption. If the imaginary part of the index is negative, we would have gain, as in a laser. One must have a population inversion to have a negative imaginary index in the steady state, and so one must have at least three levels for a typical laser system.

2.3 Slowly varying envelope approximation

For optical frequencies, it is often very convenient to separate the electric field into a rapidly oscillating part, with a slowly varying envelope function, of the form

$$E = A(t)e^{i(kx-\omega t+\phi)} \tag{2.49}$$

In what follows, we assume that the frequencies involved in the Fourier expansion of the slowly varying amplitude $A(t)$ are small compared to the center, or carrier, frequency ω. Formally, this means we take $\omega A \gg \dot{A}$. We ignore any spatial dependence of the envelope for now. Let us substitute our form for the field back into the wave equation. To do this, we need the following results:

$$\frac{\partial E}{\partial z} = ikEe^{i(kx-\omega t+\phi)} \tag{2.50}$$

$$\frac{\partial^2 E}{\partial z^2} = -k^2Ee^{i(kx-\omega t+\phi)} \tag{2.51}$$

$$\frac{\partial E}{\partial t} = (\dot{A} - i(\omega + \dot{\phi})A)e^{i(kx-\omega t+\phi)} \tag{2.52}$$

$$\frac{\partial^2 E}{\partial t^2} = (\ddot{A} - 2i(\omega + \dot{\phi})\dot{A} - (\omega + \dot{\phi})^2 A - i\ddot{\phi}A)e^{i(kx-\omega t+\phi)} \tag{2.53}$$

We must also decompose the polarization into rapidly and slowly varying components

$$P = \mathcal{P}(t)e^{i(kx-\omega t+\phi)} \tag{2.54}$$

One then has

$$\frac{\partial^2 P}{\partial t^2} = (\ddot{\mathcal{P}} - 2i(\omega + \dot{\phi})\dot{\mathcal{P}} - (\omega + \dot{\phi})^2\mathcal{P} - i\ddot{\phi}\mathcal{P})e^{i(kx-\omega t+\phi)} \tag{2.55}$$

The wave equation then becomes

$$(\Omega^2 - (\omega + \dot{\phi})^2)A - i\omega\frac{\sigma}{\varepsilon_0} - 2i\omega\dot{A} = \frac{\omega^2}{\varepsilon_0}\mathcal{P} \tag{2.56}$$

where we have used the slowly varying envelope approximation, $\ddot{O} \ll \omega\dot{O}$ and $\dot{O} \ll \omega O$, for $O = \phi$, A, or $\mathcal{P}$ and used $\Omega = ck$. Another approximation is, ignore $\dot{\phi}$ compared to $\omega + \Omega$, and take $\Omega \sim \omega$, which allows us to write

$$\begin{aligned}\Omega^2 - (\omega + \dot{\phi})^2 &= (\Omega - (\omega + \dot{\phi}))(\Omega + \omega + \dot{\phi}) \\ &\sim 2\omega(\Omega - \omega - \dot{\phi})\end{aligned} \tag{2.57}$$

On equating real and imaginary parts then, we find

$$\dot{A} = -\kappa A - \frac{1}{2}\frac{\omega}{\varepsilon_0}\Im(\mathcal{P}) \tag{2.58}$$

$$\dot{\phi} = \Omega - \omega - \frac{1}{2A}\frac{\omega}{\varepsilon_0}\Im(\mathcal{P}) \tag{2.59}$$

where in terms of our fictional conductivity we have $\kappa = \omega/2Q$, where Q is the quality factor $\varepsilon_0\omega/\sigma$. Usually, this loss rate results from a lossy mirror in a cavity. A photon has a probability of $T = 1 - R$ of leaving the mirror when it hits, where R is the reflectivity and T is the transmission of the mirror. The rate at which the photon interacts with the mirror is $1/\tau$, with $c\tau = 2L$. Overall this leads to $\kappa = (1 - R)c/2L$. For a linear material, with an imaginary susceptibility $\chi = \chi_R + i\chi_I$, we have

$$\dot{A} = -\kappa A - \frac{1}{2}\omega\chi_I A \tag{2.60}$$

$$\Delta\omega = \dot{\phi} = \Omega - \omega - \frac{1}{2}\omega\chi_R \tag{2.61}$$

Here, we see that the sign of the imaginary part of the susceptibility determines whether the interaction with the medium results in gain or loss, and that there is a frequency shift of the field that depends on the real part of the susceptibility. Recall that the relation to the index of refraction is $n = \sqrt{1 + \chi} \sim 1 + \chi_R/2 + i\chi_I/2$, where we assume a dilute medium ($n \sim 1$).

It is sometimes more convenient to consider a complex field variable

$$X = Ae^{i\phi} \tag{2.62}$$

which satisfies the equation

$$\dot{X} = -\kappa X - \frac{\omega}{\varepsilon_0}P \tag{2.63}$$

2.4 Lorentz oscillator model of the atom

The simplest model of the atom, hydrogen say, is that of an electron and proton connected by a spring. At first, this seems ridiculous until we realize that due to the Coulomb interaction the proton and electron form a bound state. For any bound state, there must be a minimum in the potential as shown below.

Near that minimum, we can expand the potential in a Taylor series,

$$V(x) = V(x_0) + \frac{dV}{dx}|_{x=x_0}(x - x_0) + \frac{1}{2}\frac{d^2V}{dx^2}|_{x=x_0}(x - x_0)^2 + \cdots \tag{2.64}$$

We are always free to define $V(x_0) = 0$ and as it is a minimum, we have $\frac{dV}{dx}|_{x=x_0}(x - x_0) = 0$, which leaves us with

$$V(x) \sim \frac{1}{2}\frac{d^2V}{dx^2}|_{x=x_0}(x - x_0)^2 = \frac{1}{2}k(x - x_0)^2 \tag{2.65}$$

where k is an effective spring constant, not to be confused with the wave number $k = 2\pi/\lambda$, Coulomb's constant $k = 1/4\pi\varepsilon_0$, or Boltzmann's constant. Can you imagine the turmoil if we are to study potassium (K) ions (interacting with a Coulomb force) with light of wave number k, at thermal equilibrium at temperature T. Anyway, I digress. The equation of motion of the oscillator driven by a cosinusoidal electric field with amplitude E and angular frequency ω would be given by

$$\ddot{x} = -\omega_0^2 x - \Gamma\dot{x} + \frac{eE}{m}\cos(\omega t) \tag{2.66}$$

where we have introduced a frictional force that is proportional to the velocity. Later we will relate that damping force to spontaneous emission. As we have a linear oscillator, it will respond in the steady state at the frequency it is driven at, with some amplitude and phase. We then assume a steady state solution of the form

$$x = X_0\cos(\omega t + \theta) \tag{2.67}$$

substituting this into the equation of motion we find

$$(\omega_0^2 - \omega^2)X\cos(\omega t + \theta) - \omega\Gamma X\sin(\omega t + \theta) = -\frac{eE}{m}\cos(\omega t) \tag{2.68}$$

We use trig identities of the form $\cos(\omega t + \theta) = \cos(\omega t)\cos(\theta) - \sin(\omega t)\sin(\theta)$ and a similar one for $\sin(\omega t + \theta)$, and equate sin and cos terms and we find

$$\theta = \tan^{-1}\frac{\omega\Gamma}{\omega_0^2 - \omega^2} \tag{2.69}$$

and

$$X_0 = \frac{-eE/m}{\sqrt{(\omega_0^2 - \omega^2)^2 + \omega^2\Gamma^2}} \tag{2.70}$$

Here, we see the familiar results that the amplitude of the response is maximized on resonance ($\omega = \omega_0$), and that on resonance the response is $\pi/2$ out of phase with the driving field. We can write the response of the oscillator as

$$\begin{aligned} x(t) &= X_0\cos(\omega t)\cos(\phi) - X_0\sin(\omega t)\sin(\phi) \\ &= X_0(U\cos(\omega t) - V\sin(\omega t)) \end{aligned} \tag{2.71}$$

with $U = \cos(\phi)$ the relative amplitude in phase with the driving field and $V = \sin\phi$ the relative amplitude in quadrature ($\pi/2$ out of phase) with the driving field. There is a slightly easier method to obtain these results, were we use a complex representation. The driving field has a time dependence $\cos(\omega t) = Re(e^{i\omega t})$. As this is a linear system, we may take the driving field to be complex, find the response, and then take the real part at the end. Thus, we need to solve

$$\ddot{x} = -\omega_0^2 x - \Gamma\dot{x} + \frac{eE}{m}e^{i\omega t} \tag{2.72}$$

with steady state response

$$x(t) = \tilde{X}_0 e^{i\omega t} \tag{2.73}$$

This leads to

$$\tilde{X}_0 = \frac{-eE/m}{(\omega_0^2 - \omega^2) - i\omega\Gamma} \tag{2.74}$$

which leads to

$$x(t) = X_0 \cos(\omega t + \theta) \tag{2.75}$$

with the same X_0 and θ we found earlier. This gives us an expression for the steady state polarization of the medium via

$$\begin{aligned} P &= Nex(t) \\ &= NeX_0 \cos(\omega t + \theta) \\ &= \varepsilon_0 \chi E(t) \end{aligned} \tag{2.76}$$

which leads to an expression for the susceptibility

$$\chi = \frac{Ne^2/m\varepsilon_0}{(\omega_0^2 - \omega^2) - i\omega\Gamma} \tag{2.77}$$

which is a complex expression with real and imaginary parts

$$\chi_R = \frac{Ne^2/m\varepsilon_0}{(\omega_0^2 - \omega^2)^2 + \Gamma^2\omega^2}(\omega_0^2 - \omega^2) \tag{2.78}$$

$$\chi_I = \frac{Ne^2/m\varepsilon_0}{(\omega_0^2 - \omega^2)^2 + \Gamma^2\omega^2}\omega\Gamma \tag{2.79}$$

For a dilute medium, we then have $n_R - 1 = \chi_R/2$ and $n_I = \chi_I/2$.

The power dissipated by the oscillator is Power $= Fv = F\dot{x}$ which leads to

$$\text{Power} = \frac{e^2E^2\omega/m}{\sqrt{(\omega_0^2 - \omega^2)^2 + \omega^2\Gamma^2}} \cos(\omega t)\cos(\omega t + \beta) \tag{2.80}$$

where we have taken $\beta = \theta - \pi/2$ which is zero on resonance. A time average of this expression yields

$$\begin{aligned} \text{Power} &= \frac{e^2E^2\omega/2m}{\sqrt{(\omega_0^2 - \omega^2)^2 + \omega^2\Gamma^2}} \cos(\beta) \\ &= \frac{e^2E^2\omega/2m}{(\omega_0^2 - \omega^2)^2 + \omega^2\Gamma^2} \\ &= \omega\varepsilon_0 E^2 n_I \end{aligned} \tag{2.81}$$

This is consistent with our earlier observation that the imaginary part of the index of refraction is responsible for changes in the amplitude of a wave as it propagates. Here, we are putting energy into the oscillator via the applied field, which is then dissipated as shown above. What is the physical mechanism for this? Well the electron is not connected to the proton by a spring, which has some velocity dependent friction associated with it! How would a real atom dissipate energy? The only mechanism we know of in a real, quantum mechanical theory, is spontaneous emission. When we examine the quantum theory of an atom interacting with an applied field, we will find that for weak excitation, the results above are correct from a quantum mechanical point of view if we identify Γ as half the spontaneous emission rate.

How about the dynamical response of the oscillator? The time dependence of the approach to the steady state is given by equation (2.66). We will make the decomposition

$$x(t) = X_0(U\cos(\omega t) - V\sin(\omega t)) \tag{2.82}$$

and substitute this ansatz into equation (2.66), which, upon equating coefficients of sin and cos, leads to equations

$$\dot{U} = \Delta V - \Gamma U \tag{2.83}$$

$$\dot{V} = -\Delta U - \Gamma V + \Omega_{cl} \tag{2.84}$$

where we define $\Delta = \omega_0 - \omega$, $\Omega_{cl} = eE/2m\omega X_0$ we will refer to as a classical Rabi frequency, and we have made liberal use of the slowly varying amplitude approximation, for example, $\omega U \gg \dot{U}$ and $\omega V \gg \dot{V}$. In terms of these quantities, the polarization is

$$\begin{aligned} P &= NeX_0(U\cos(\omega t) - V\sin(\omega t)) \\ &= \frac{NeX_0}{2}((U + iV)e^{i\omega t} + (U - iV)e^{-i\omega t}) \end{aligned} \tag{2.85}$$

which in a complex representation leads to

$$\chi_R = \frac{NeX_0}{2}\frac{U}{E} \tag{2.86}$$

$$\chi_I = \frac{NeX_0}{2}\frac{V}{E} \tag{2.87}$$

Then, we can write down the full set of coupled equations for the field and the polarization

$$\dot{U} = \Delta V - \Gamma U \tag{2.88}$$

$$\dot{V} = -\Delta U - \Gamma V + \Omega_{cl} \tag{2.89}$$

$$\dot{A} = -\kappa A - \frac{1}{2}\omega V \tag{2.90}$$

$$\dot{\phi} = \Omega - \omega - \frac{1}{2}\omega U \tag{2.91}$$

References

[1] Bromley D A and Greiner W 1998 *Classical Electrodynamics* Classical Theoretical Physics (New York: Springer)

[2] Cohen-Tannoudji C, Dupont-Roc J and Grynberg G 1998 *Atom-Photon Interactions: Basic Processes and Applications* (New York: Wiley)

[3] Cohen-Tannoudji C, Dupont-Roc J and Grynberg G 1989 *Photons and Atoms: Introduction to Quantum Electrodynamics* (New York: Wiley)

[4] Griffiths D J 1999 *Introduction to Electrodynamics* (Englewood Cliffs, NJ: Prentice-Hall)

[5] Jackson J D 1975 *Classical Electrodynamics* (New York: Wiley)

[6] Landau L D, Bell J S, Kearsley M J, Pitaevskii L P, Lifshitz E M and Sykes J B 2013 *Electrodynamics of Continuous Media* (Amsterdam: Elsevier)

Chapter 3

QM review

We give a brief review of quantum mechanics, with attention to the usage of Hilbert spaces and Dirac notation. Standard references include [1–7]

3.1 Wave mechanics

In all of your previous quantum mechanics you have dealt with the wave function $\Psi(\vec{x}, t)$. This function describes the state of a quantum system. The time evolution of this function is given by the Schrödinger equation,

$$\imath\hbar\frac{\partial\Psi(\vec{x}, t)}{\partial t} = H\Psi(\vec{x}, t). \tag{3.1}$$

The Hamiltonian is generally the sum of the kinetic energy and potential energy operators

$$H = p^2/2m + V(\vec{x}, t). \tag{3.2}$$

In the position representation, the momentum operator is given by $p_x = -\imath\hbar\partial/\partial x$ in one dimension, or $\vec{p} = -\imath\hbar\vec{\nabla}$ in three spatial dimensions, which makes the Schrödinger equation take the familiar form

$$\imath\hbar\frac{\partial\Psi(\vec{x}, t)}{\partial t} = -\frac{\hbar^2}{2m}\nabla^2\Psi(\vec{x}, t) + V(\vec{x}, t)\Psi(\vec{x}, t). \tag{3.3}$$

In the special case that the potential energy depends only on position and not time, we may use separation of variables to reduce the Schrödinger equation, a partial differential equation in time and space, to two equations involving spatial dependence and temporal dependence. We use the ansatz

$$\Psi(\vec{x}, t) = \psi(\vec{x})F(t). \tag{3.4}$$

doi:10.1088/978-0-7503-1713-9ch3

Substituting this into the Schrödinger equation and dividing it by $\Psi(\vec{x}, t)$ yields

$$\frac{\imath\hbar}{F(t)}\frac{\partial F(t)}{\partial t} = -\frac{\hbar^2}{2m\psi(\vec{x})}\nabla^2\psi(\vec{x}) + V(\vec{x}) = E, \tag{3.5}$$

where E is a constant of separation. The function $F(t)$ is then given by

$$F(t) = e^{-\imath Et/\hbar}. \tag{3.6}$$

We gave ignored an overall constant factor, as the full wavefunction $\Psi(\vec{x}, t)$ must be normalized, and we choose to normalize the spatial and temporal parts separately. As written, $F^*(t)F(t) = 1$, and ultimately we normalize the spatial part of the wave function. The function $\psi(\vec{x})$ then satisfies the time-independent Schrödinger equation

$$-\frac{\hbar^2}{2m}\nabla^2\psi(\vec{x}) + V(\vec{x}, t)\psi(\vec{x}) = H\psi(\vec{x}) = E\psi(\vec{x}). \tag{3.7}$$

As the left-hand side of the equation is the sum of the potential and kinetic energy operators, we associate the constant E with the total energy of the system.

Most introductory work on quantum mechanics concerns specifying a potential energy $V(\vec{x})$, and solving the above equation for the allowed energies E_n and wavefunctions $\psi_n(\vec{x})$. The subscript n is used to indicate that there is a set of possible solutions, and not just one. It is important to realize that the time-independent Schrödinger equation $H\psi_n(\vec{x}) = E_n\psi_n(\vec{x})$ is an eigenvalue equation for *BOTH* $\psi_n(\vec{x})$ as well as E_n. Generally in quantum mechanics, we deal with Hamiltonians that are Hermitian operators, so the eigenenergies E_n are real, and the eigenfunctions $\psi_n(\vec{x})$ are orthogonal, and can be made orthonormal

$$\int_{\infty}^{\infty} \psi_n^*(\vec{x})\psi_m(\vec{x})d\vec{x} = \delta_{n,m}, \tag{3.8}$$

where we have normalized the individual wave functions via

$$\int_{\infty}^{\infty} \psi_n^*(\vec{x})\psi_n(\vec{x})d\vec{x} = 1. \tag{3.9}$$

Recall that the concept of a Hermitian operator in infinite dimensions, as we have here (an infinite number of possible positions $\vec{x}$) means that the operator is self-adjoint, and that the solutions satisfy certain boundary conditions. As one choice of appropriate boundary conditions is that the wave functions and their derivatives vanish at $\pm\infty$, this is automatically satisfied for any wavefunction that can be normalized. One can look at a matrix and quickly determine whether it is Hermitian or not; for a differential operator we can only tell whether it is self-adjoint or not. But with appropriate boundary conditions, that we require in quantum mechanics, the two concepts are essentially one and the same.

The individual energy eigenstates are 'stationary' as $|\Psi(\vec{x}, t)|^2 = |\psi(\vec{x})|^2$ is independent of time. We recall the Born interpretation that the probability of finding the particle between x and $x + dx$ is given by $P(x)dx = |\psi(\vec{x})|^2dx$, so the stationarity is that of the spatial probability distribution. As the Schrödinger equation is linear, the principle of superposition yields the most general form of the solution

$$\Psi(\vec{x}, t) = \sum_{n=1}^{\infty} C_n \psi_n(\vec{x}) e^{-\iota\omega_n t} \tag{3.10}$$

where we have used the Bohr condition $E_n = \hbar\omega_n$. It can be shown that

$$\sum_{n=1}^{\infty} |C_n|^2 = 1 \tag{3.11}$$

From this follows, the interpretation of $|C_n|^2 = C_n C_n^*$ as a probability of finding E_n as a result of measuring the energy of the system. After such a measurement, the wavefunction is in the state $\psi_n(\vec{x})$, the 'collapse' of the wavefunction.

Note that the most general state is *NOT* stationary! The probability of obtaining a certain value of $E = E_n$ is independent of time, but other probabilities, such as the probability of finding the particle at a certain position x is not. In fact, for a simple superposition such as $\psi(\vec{x}) = \sqrt{1/2}(\psi_1 + \psi_3)$, we have (in 1D)

$$P(x)dx = |\Psi|^2 dx = (1/2)\left(|\psi_1|^2 + |\psi_3|^2 + 2Re\left(\psi_1^* \psi_3 e^{-\iota(\omega_3 - \omega_1)t}\right)\right)dx \tag{3.12}$$

In a very real sense the system is simultaneously in all eigenstates with a nonzero probability amplitude C_n. Quantum mechanics is certainly statistical in nature, in that we cannot tell before a measurement of energy which eigenvalue E_n we will obtain, but only the probability of each possible result. But it is more than statistical. We cannot say that the system has a definite energy, but we have a certain lack of knowledge about which one it is. The work of Bell and others has shown us that the implications of quantum mechanics are deeper than that.

We recall that once the wavefunction has been found, the information about a system in that state, or more precisely the average value of a measurement of some observable O on an ensemble of identically prepared systems, is given by

$$\langle O \rangle = \int_{-\infty}^{\infty} \Psi^*(\vec{x}, t) O \Psi(\vec{x}, t) d\vec{x} \tag{3.13}$$

We have again assumed that the individual eigenstates have been normalized. It is also convenient to consider the spread, or 'uncertainty' of a set of measurements,

$$(\Delta O)^2 = \langle (O - \langle O \rangle)^2 \rangle = \langle O^2 \rangle - \langle O \rangle^2 \tag{3.14}$$

At this point, we also note that representing physical observables by Hermitian differential operators preserves the feature that one notices in the lab in the quantum realm; that is that order of measurement matters! If I make a position measurement and then a momentum measurement, I will in general get different results than if I make a momentum measurement first and then a position measurement. Consider the commutator of the operators x and p, acting on a wavefunction ψ,

$$\begin{aligned}[x, p]\psi &\equiv (xp - px)\psi \\ &= -\imath\hbar\left(x\frac{\partial}{\partial x} - \frac{\partial}{\partial x}x\right)\psi \\ &= -\imath\hbar x\frac{\partial\psi}{\partial x} + \imath\hbar\frac{\partial(x\psi)}{\partial x} \\ &= \imath\hbar\psi\end{aligned} \tag{3.15}$$

Here, we see that differential operators can take into account that 'order matters', as the operators representing physical quantities do not commute. There are other choices of mathematical objects that do not in general commute, such as matrices, which we shall also use later on. If x and p were numbers, then obviously $[x, p] = 0$, but in the quantum case we find that $[x, p] = \imath\hbar$ which is *NOT* zero, unless we take the classical limit that $\hbar \to 0$. This non-commutativity leads directly to the uncertainty principle for two Hermitian operators A and B,

$$\Delta A\Delta B \geqslant \frac{1}{-}|\langle -i[A, B]\rangle|^2 \tag{3.16}$$

which is generally state dependent. For position and momentum becomes

$$\Delta x\Delta p \geqslant \hbar/2 \tag{3.17}$$

3.2 Dirac notation

We now move to a more abstract view of the state of a quantum system. We will represent the state of a quantum system by a ray in a linear vector space. For a review of linear vector spaces and their application to quantum mechanics, we refer you to Shankar's texts, and Arfken and Weber's texts. Typically, the ray that represents the quantum state is notated as $|\Psi(\vec{x}, t)\rangle$. This is known as a 'ket'. This is also known as the state vector, and may have a representation as a column of numbers, in a given basis (or in simpler terms an orthonormal set of vectors that forms a coordinate system). But it is an abstract property of a quantum system, independent of basis. Think about the electric field at some point in space, it exists independent of our description of it; our representation of it (the three components) may vary in different coordinate systems. There is also a corresponding vector referred to as the 'bra', $\langle\Psi(\vec{x}, t)|$, which usually has a representation as a row vector, whose elements are the complex conjugates of the components of the column vector 'ket'. The bra is given by

$$\langle\psi(x)| = (\psi(x_1)\psi(x_2)\psi(x_3)\ldots) \tag{3.18}$$

How do we connect these objects? What about $\psi(x)$, the wave function. Is there any way to represent a function as a vector? The answer is yes, we merely list the value of the function at all points x_i. This vector is infinite in length of course, and so it less than useful other than in the abstract

$$|\psi(x)\rangle = \begin{pmatrix} \psi(x_1) \\ \psi(x_2) \\ \psi(x_3) \\ \vdots \end{pmatrix} \tag{3.19}$$

If we were just to sample the wave function at certain equally spaced points, we would have a function 'vector' such as you have created in excel or another spreadsheet to make a plot of the function. In the abstract though, we wish to consider sampling at every point, which does make for a long vector. The bra is given by

$$\langle\psi(x)| = (\psi^*(x_1), \psi^*(x_2), \psi^*(x_3)\ldots) \tag{3.20}$$

The inner product $\langle\psi(x)|\phi(x)\rangle$ is an infinite sum that in general does not converge. The sum is generally infinite due to the infinite length of the vectors involved, independent of the nature of the functions involved. If we think of sampling the wave function over an interval L at N points separated by a distance dx, the size of the inner product scales as N. If we multiply the usual inner product by $dx = L/N$, and take the limit of the sum over into an integral we find

$$\langle\psi(x)|\phi(x)\rangle = \int \psi(x)^*\phi(x)dx \tag{3.21}$$

This defines the inner product on our vector space, called a Hilbert space. It is linear and has other nice properties. We note that we choose *this* inner product over others as its utility for doing quantum mechanics. The probability for finding the particle between x and $x + dx$ is given by $P(x)dx = \psi(x)^*\psi(x)dx$. The requirement that the probability of finding the particle anywhere being one results in

$$\langle\psi(x)|\psi(x)\rangle = \int_{-\infty}^{\infty} \psi(x)^*\psi(x)dx = 1 \tag{3.22}$$

which is referred to as the normalization condition. As $A\psi(x)$ is a solution to the Schrödinger equation if $\psi(x)$ is, the wave function must be normalized after finding the form of the solution. But why do we need to consider $\psi(x)^*$? Wave functions are in general complex of course, but why is the inner product defined in this way?

We could have defined the square of the 'length' of the quantum state vector in the usual sense of the dot product of the vector with itself. This would not always generate a real number, or even one with a positive real part. If, for example

$$|\psi\rangle = \begin{pmatrix} i \\ 1 \\ i \end{pmatrix} \tag{3.23}$$

then the 'length' of the quantum state would be $-1 + 1 - 1 = -1$. It is more convenient to generalize the dot product into an inner product. To do this, we define a bra

$$\langle\psi| = (C_1^* \;\; C_2^* \;\; C_3^*) \tag{3.24}$$

We then define the inner product as the matrix multiplication of the bra and ket

$$\langle\psi||\psi\rangle \equiv \langle\psi|\psi\rangle \tag{3.25}$$

$$= (C_1^* \;\; C_2^* \;\; C_3^*)\begin{pmatrix} C_1 \\ C_2 \\ C_3 \end{pmatrix} \tag{3.26}$$

$$= |C_1|^2 + |C_2|^2 + |C_3|^2 \tag{3.27}$$

This general definition of an inner product always results in a positive definite quantity. For a normalized quantum state, where the total probability of being in any state is one, then this inner product yields unity, as $|C_1|^2 + |C_2|^2 + |C_3|^2 = P_1 + P_2 + P_3 = 1$. One can take the inner product $\langle\psi|\phi\rangle$ which will not generally yield a real positive number. The names bra and ket stem from the fact that the inner product $\langle\psi|\phi\rangle$ is formed inside a 'bracket'. This notation is due to Paul Dirac, and is called Dirac notation, and generalizes to a state with N levels

$$|\psi\rangle = \begin{pmatrix} C_1 \\ C_2 \\ C_3 \\ \vdots \end{pmatrix} \tag{3.28}$$

3.3 Representations and pictures

So far, we have only dealt with what is called the position representation, where we utilize

$$\begin{aligned} \hat{p}_x &= -i\hbar\frac{\partial}{\partial x} \\ \hat{x} &= x \end{aligned} \tag{3.29}$$

This representation is obtained by considering the matrix representation of the position and momentum operators in a basis of eigenstates of the position operator. We are interested in states that satisfy $\hat{x}|x\rangle = x|x\rangle$. These states form a complete set, as they are eigenvectors of a Hermitian operator,

$$\int dx|x\rangle\langle x| = 1 \tag{3.30}$$

This leads us to the normalization of the position eigenstates as

$$\begin{aligned}\langle x|x'\rangle &= \int dx''\langle x|x''\rangle\langle x|x'\rangle \\ &= \int dx'\delta(x - x'')\langle x''|x'\rangle\end{aligned} \tag{3.31}$$

and we see we have $\langle x|x'\rangle = \delta(x - x')$ via the definition of the Dirac delta function. This is zero for $x \neq x'$, but infinite for $x = x'$. Recall that the Dirac delta function is really a distribution, and makes much more sense when it lives inside an integral, but we shall have occasion to use it outside of integration. At this point, we can make a definite connection between the ket $|\psi\rangle$ and our old friend the wave function $\psi(x)$. Let us use the completeness relationship to understand the connection.

$$\begin{aligned}|\psi\rangle &= \int dx|x\rangle\langle x|\psi\rangle \\ &= \int dx\langle x|\psi\rangle|x\rangle \\ &= \int dxC(x)|x\rangle\end{aligned} \tag{3.32}$$

This is merely a way to express that we can write the state vector as a linear superposition of position eigenstates, with probability amplitude $C(x) = \langle x|\psi\rangle$. Recall from wave mechanics the requirement of normalization of the wave function

$$\int_{-\infty}^{\infty} \psi^*(x)\psi(x)dx = \int_{-\infty}^{\infty} P(x)dx = 1 \tag{3.33}$$

this allows us to interpret $P(x)dx = \psi^*(x)\psi(x)dx$ as a probability density for finding the particle in a region of space $x \rightarrow x + dx$. In the Dirac formalism we have, using the completeness relationship,

$$\begin{aligned}\langle\psi|\psi\rangle &= \int dx\langle\psi|x\rangle\langle x|\psi\rangle \\ &= \int C^*(x)C(x)dx \\ &= 1\end{aligned} \tag{3.34}$$

where we have normalized the 'length' of the state vector to unity. Requiring this normalization for the expansion equation (3.32) leads us to $P(x) = |C(x)|^2$, or that the expansion coefficient is just $C(x) = \psi(x)$.

Using the eigenstates of the position operator yields what is the position representation. The matrix elements of the position operator are

$$\begin{aligned}\langle x'|\hat{x}|x\rangle &= \int dx''\langle x'|x''\rangle\langle x''|\hat{x}|x\rangle \\ &= \int dx''x\langle x'|x''\rangle\langle x''|x\rangle \\ &= \int dx''\delta(x' - x'')\delta(x - x'') \\ &= x\delta(x - x')\end{aligned} \tag{3.35}$$

Consider now the expectation value of an operator defined by

$$\begin{aligned}\langle\hat{O}\rangle &\equiv \langle\psi|\hat{O}|\psi\rangle \\ &= \int dx \int dx' \langle\psi|x'\rangle\langle x'|\hat{O}|x\rangle\langle x|\psi\rangle \\ &= \int dx \int dx' \psi^*(x') O_{x,x'} \psi(x)\end{aligned} \tag{3.36}$$

with the matrix element of the operator $\hat{O}$ between two position eigenstates defined as $\langle x'|\hat{O}|x\rangle \equiv O_{x,x'}$. This is just the inner product of a column vector and a row vector, with a matrix for the operator sandwiched in between. For $\hat{O} = \hat{x}^n$, we have,

$$\begin{aligned}\langle\hat{x}^n\rangle &= \int dx \int dx' \psi^*(x') x^n \delta(x - x')\psi(x) \\ &= \int dx \psi^*(x) x^n \delta(x - x')\psi(x) \\ &= \int dx x^n P(x)\end{aligned} \tag{3.37}$$

which again just tells us that $P(x) = |\psi(x)|^2$ is a probability distribution with respect to x. And in this representation based on eigenstates of the position operator, we can always replace the operator $\hat{x}$ with the scalar x. The general recipe equation (3.36) holds for all operators.

The matrix elements of $\hat{K} = -i$ times the derivative operator in the position representation are

$$\langle x|\hat{K}|x'\rangle = -\imath\delta'(x - x') \tag{3.38}$$

where $\delta'(x - x')$ is the derivative of the delta function with respect to x'. This operator is Hermitian; if we multiply it by $\hbar$ we find the momentum operator $\hat{p} = \hbar\hat{K}$. Calculating expectation values of the momentum operator using equation (3.36), we find

$$\begin{aligned}\langle\hat{p}^n\rangle &= \int dx \int dx' \psi^*(x')(-\imath\hbar)^n \delta'^n(x - x')\psi(x) \\ &= (-\imath\hbar)^n \int dx \psi^*(x)\delta'^n(x - x')\psi(x) \\ &= (-\imath\hbar)^n \int dx \psi^*(x) \frac{d^n\psi(x)}{dx^n}\end{aligned} \tag{3.39}$$

So as long as we remember to put the momentum operator *between* the wave function and its complex conjugate in the integral, we may always use $\hat{p} \to -\imath\hbar d/dx$.

Now why should position eigenstates occupy such a special place in our theory? We should be able to use any complete set of states to expand our state vector, and the eigenstates of any Hermitian operator yield a complete set of normalizable eigenstates. What about using the position eigenstates defined by

$$\hat{p}|p\rangle = p|p\rangle \tag{3.40}$$

where the momentum eigenstates form a complete basis satisfying

$$\int dp|p\rangle\langle p| = 1 \tag{3.41}$$

Going through a similar analysis we find that for these states we have

$$\langle\hat{p}^n\rangle = \int_{-\infty}^{\infty} dpp^n\Phi^*(p)\Phi(p) \tag{3.42}$$

$$\langle\hat{x}^n\rangle = (\imath\hbar)^n \int_{-\infty}^{\infty} dp\Phi^*(p)\frac{d^n\Phi(p)}{dp^n} \tag{3.43}$$

where we have defined the momentum space wave function $\Phi(p) = \langle p|\psi\rangle$.

The connection between these two representations is found by considering the character of the momentum eigenstates in the position representation, $\langle x|p\rangle$. Inserting the appropriate completeness relationships, we find

$$\psi_p(x) = \langle x|p\rangle = \frac{1}{\sqrt{2\pi}}\exp(ipx/\hbar) \tag{3.44}$$

This is just a plane wave, a particle with a definite momentum $p = \hbar k$, definite energy (for a free particle) of $p^2/2m$, and a completely indeterminate position, $P(x) =$ constant. Using this, we find

$$\begin{aligned}\Phi(p) &= \langle p|\psi\rangle \\ &= \int dx\langle p|x\rangle\langle x|\psi\rangle \\ &= \frac{1}{\sqrt{2\pi}}\int dx e^{\imath px/\hbar}\psi(x)\end{aligned} \tag{3.45}$$

and so the position and momentum space wave functions form a Fourier pair. As Fourier pairs satisfy the relationship $\Delta x\Delta k \geqslant 1/2$, we have the uncertainty principle showing up. Multiply by $\hbar$ to obtain $\Delta x\Delta p \geqslant \hbar/2$, where we have $p = \hbar k$. Let us pause and examine the cause of the 2π; we have played a little fast and loose with it so far. Recall that the plane wave function cannot be normalized in the ordinary sense. It has a mod squared of unity, and the integral over all x would lead to infinity. It is just as problematic as a Dirac delta function, as well it should be, as they form a Fourier pair! We require the orthonormality of the momentum space wave functions to be given in terms of a Dirac delta function, as we did for position eigenstates, and we have

$$\langle k|k'\rangle = \delta(k - k') = \int_x \langle k|x\rangle\langle x|k'\rangle = \frac{1}{2\pi}\int dx e^{\imath(k-k')x} \tag{3.46}$$

where the 2π factor comes from our use of an integral representation of the Dirac delta function.

Why have you not seen the momentum space representation before? It is just as valid. The main reason is that we usually deal with potentials that are functions of position rather than momentum, and so we have an easier time solving differential equations in the position representation. Let us consider first the harmonic oscillator. We have $\hat{H} = \hat{p}^2/2m + k\hat{x}^2/2$. In the position representation, the time independent Schrödinger equation becomes

$$\frac{-\hbar^2}{2m}\frac{d^2\psi_n}{dx^2} + \frac{kx^2\psi_n}{2} = E_n\psi_n \tag{3.47}$$

which can be solved to find energy levels given by $E_n = (n + 1/2)\hbar\omega$ with $\omega^2 = k/m$; and wave functions that are a Gaussian multiplied by a Hermite polynomial of order n. How would this look in the momentum space representation?

$$\frac{p^2\Phi_l}{2m} + \frac{k}{2}\frac{d^2\Phi_l}{dp^2} = E_l\Phi_l \tag{3.48}$$

They are basically the same form, a term quadratic in the variable natural to the representation, and a second derivative with respect to that same variable. Obviously, something is very special about the simple harmonic oscillator! We get the same type of wave function independent of our choice of special variable.

What about other systems? Consider a simple 1D Coulomb type potential, $V(x) = \eta/x$. In the position representation, we have

$$\frac{-\hbar^2}{2m}\frac{d^2\psi_n}{dx^2} + \frac{\eta\psi_n}{x} = E_n\psi_n \tag{3.49}$$

Now this may or may not have pretty solutions, but it does not seem too bad for now. What about the momentum representation?

$$\frac{p^2\Phi_l}{2m} + \frac{\eta}{\partial/\partial p}\Phi = E_l\Phi_l \tag{3.50}$$

Well it is a very different form in this representation, and it is not quite clear how to interpret the derivative in the denominator acting on Φ. Obviously, there is something very special about the harmonic oscillator which we will discuss in more detail later.

3.4 Pictures

Gee, in quantum mechanics, we also have pictures. What is a picture? It is a choice of where to put the time dependence in our equations. Usually, one uses the Schrödinger picture where the wave function evolves in time via

$$\imath\hbar\frac{\partial|\psi(x, t)\rangle}{\partial t} = \hat{H}\psi(x, t) \tag{3.51}$$

and operators that are independent of time like

$$\hat{H} = \hat{p}^2/2m + V(\hat{x}) \tag{3.52}$$

$$\hat{x} \to x \tag{3.53}$$

$$\hat{p} \to -\imath\hbar\partial/\partial x \tag{3.54}$$

We can write down a formal solution to the Schrödinger equation,

$$|\psi(x, t)\rangle = \exp(-\imath\hat{H}t/\hbar)|\psi(0, t)\rangle \tag{3.55}$$

To this point in the discussion, as we are using the time independent Schrödinger equation, recall that $\hat{H}$ is independent of time, at least for now. We refer to the operator $\exp(-\imath\hat{H}t/\hbar) = U(t)$ as the propagator, as it propagates our state vector in time.

Maybe you have not run across the exponential of an operator before. It is generally well defined, consider

$$e^x = \sum_{n=0}^{\infty}\frac{x^n}{n!} = 1 + x + x^2/2 + x^3/6 + \cdots \tag{3.56}$$

So, as long as multiplication of an operator is well defined, as the multiple action of the operator on a state vector, functions of operators are also well defined. The propagator yields unitary evolution if $\hat{H}$ is Hermitian as we generally require; that is, the norm of the state function does not change under this temporal evolution.

The propagator can be used to solve the Schrödinger equation numerically, essentially using Simpson's rule for small timesteps Δt

$$|\psi(t + \Delta t)\rangle = \exp(-\imath\hat{H}t/\hbar)|\psi(t)\rangle \approx (1 - \imath\hat{H}\Delta t/\hbar)|\psi(t)\rangle \tag{3.57}$$

Fancier integration schemes may be more efficient, or stable, or both. With the use of the propagator, we can write down how the expectation value of an operator changes with time

$$\begin{aligned}\langle\hat{O}(t)\rangle &= \langle\psi(t)|\hat{O}|\psi(t)\rangle \\ &= \langle\psi(0)|e^{\imath\hat{H}t/\hbar}\hat{O}e^{-\imath\hat{H}t/\hbar}|\psi(0)\rangle\end{aligned} \tag{3.58}$$

We can use this as a starting point to define the Heisenberg picture, where we take all the time dependence to be in the operators, and none of the time dependence is in the state vector. The two pictures correspond at some initial time $t = 0$. We define the Heisenberg state vector to be $|\psi_H(t)\rangle = |\psi_H(0)\rangle = |\psi_S(0)\rangle$. To preserve our expression for the time evolution of expectation values, the things we actually measure, we must define the Heisenberg operator to be time dependent and given by

$$\hat{O}_H(t) = e^{\imath\hat{H}t/\hbar}\hat{O}e^{-\imath\hat{H}t/\hbar} \tag{3.59}$$

If we take a time derivative of the above, we find the equation of motion for operators

$$\begin{aligned}\frac{d\hat{O}_H}{dt} &= \frac{\imath}{\hbar}\hat{H}e^{\imath\hat{H}t/\hbar}\hat{O}_H e^{-\imath\hat{H}t/\hbar} \\ &+ e^{\imath\hat{H}t/\hbar}\frac{\partial\hat{O}_H}{\partial t}e^{-\imath\hat{H}t/\hbar} \\ &- \frac{\imath}{\hbar}e^{\imath\hat{H}t/\hbar}\hat{O}_H e^{-\imath\hat{H}t/\hbar}\end{aligned} \tag{3.60}$$

Using the definition of the Heisenberg operator, and the fact that the Hamiltonian commutes with itself (but not in general $\hat{O}$), we have

$$\frac{d\hat{O}_H}{dt} = \frac{\imath}{\hbar}[\hat{H}, \hat{O}_H] + \frac{\partial\hat{O}}{\partial t} \tag{3.61}$$

Taking expectation values of both sides, one finds Ehrenfest's theorem

$$\left\langle\frac{d\hat{O}_H}{dt}\right\rangle = \frac{\imath}{\hbar}[\langle\hat{H}, \hat{O}_H\rangle] + \left\langle\frac{\partial\hat{O}}{\partial t}\right\rangle \tag{3.62}$$

Usually, the term involving the partial derivative of the operator with time is zero; after all operators like $\hat{x}$ and $\hat{p}$ have no explicit time dependence. The first thing we notice is that if an operator has no explicit time dependence (time dependence even in the Schrödinger picture), if it commutes with the Hamiltonian, its expectation value does not change with time. This is the basis of conservation laws in quantum mechanics.

To further illustrate the utility of the Heisenberg picture, consider equations for the operators $\hat{x}$ and $\hat{p}$. For $\hat{x}$, we have

$$\begin{aligned}\frac{d\hat{x}}{dt} &= \frac{\imath}{\hbar}\left[\frac{\hat{p}^2}{2m}, \hat{x}\right] + \frac{\imath}{\hbar}[V(\hat{x}), \hat{x}] \\ &= \frac{\imath}{2m\hbar}(\hat{p}\hat{p}\hat{x} - \hat{x}\hat{p}\hat{p}) \\ &= \frac{\imath}{2m\hbar}(\hat{p}\hat{p}\hat{x} - \hat{p}\hat{x}\hat{p} + \hat{p}\hat{x}\hat{p} - \hat{x}\hat{p}\hat{p}) \\ &= \frac{\imath}{2m\hbar}(\hat{p}[\hat{p}, \hat{x}] - [\hat{p}, \hat{x}]\hat{p}) \\ &= \frac{\imath}{2m\hbar}(-2\imath\hbar\hat{p}) \\ &= \hat{p}/m\end{aligned} \tag{3.63}$$

where we have used the fact that $\hat{x}$ commutes with itself and $[\hat{x}, \hat{p}] = \imath\hbar$. This is reminiscent of the Newtonian relation $p = mv$. Let us now consider

$$\begin{aligned}\frac{d\hat{p}}{dt} &= \frac{\imath}{\hbar}\left[\frac{\hat{p}^2}{2m}, \hat{p}\right] + \frac{\imath}{\hbar}[V(\hat{x}), \hat{p}] \\ &= \frac{\imath}{\hbar}\left[V(\hat{x}), \frac{\partial}{\partial x}\right]\end{aligned} \tag{3.64}$$

To calculate this last commutator, we utilize a test function

$$\left[V(\hat{x}), \frac{\partial}{\partial x}\right]|\psi\rangle = \left(V(\hat{x})\frac{\partial|\psi\rangle}{\partial x} - V(\hat{x})\frac{\partial|\psi\rangle}{\partial x} - \frac{\partial V(\hat{x})}{\partial x}|\psi\rangle\right)$$
$$= -\frac{\partial V(\hat{x})}{\partial x}|\psi\rangle \tag{3.65}$$

As this holds for any test function that is well behaved, we have

$$\frac{d\hat{p}}{dt} = -\frac{\partial V(\hat{x})}{\partial x} \tag{3.66}$$

This is all that is left of Newton's second equation in quantum mechanics. Recall that this does not give us Newtonian equations as averages over quantum uncertainty. Newton says that $dp/dt = -dV/dx$. If we take an ensemble average over both sides that would result in

$$\frac{d\langle p_{\text{Newton}}\rangle}{dt} = -\frac{\partial\langle V(x)\rangle}{dx} \tag{3.67}$$

where Ehrenfest's theorem tells us that

$$\frac{d\langle\hat{p}\rangle}{dt} = -\left\langle\frac{\partial V(\hat{x})}{\partial x}\right\rangle \tag{3.68}$$

This is in general different, as taking a derivative and an ensemble average do not commute.

Finally, there is one final picture we would like to consider, the interaction picture. It is used often in quantum optics as well as particle physics. Here, there is time dependence in *both* the state vector and the operators. At first sight this seems to be more complicated, letting both things change. But we will see that there is a utility for this, otherwise we would not consider it! In many cases of interest, we will have a quantum system with a bare Hamiltonian $\hat{H}_0$ which will have a set of known eigenstates and energies. It may be that of a simple harmonic oscillator, a hydrogen atom, or some other nice system. We will then take this system and apply a time dependent interaction, which we shall see causes transitions, $\hat{H}_I$. We will put the time dependence arising from the bare Hamiltonian into the evolution of the operators, and the time dependence arising from the interaction into the state vectors evolution. Using the expression for the evolution of expectation values

$$\begin{aligned}\langle\hat{O}(t)\rangle &= \langle\psi(t)|\hat{O}|\psi(t)\rangle \\ &= \langle\psi(0)|e^{\imath\hat{H}t/\hbar}\hat{O}e^{-\imath\hat{H}t/\hbar}|\psi(0)\rangle \\ &= \langle\psi(0)|e^{\imath\hat{H}_I t/\hbar}e^{\imath\hat{H}_0 t/\hbar}\hat{O}e^{-\imath\hat{H}_0 t/\hbar}e^{\imath\hat{H}_I t/\hbar}|\psi(0)\rangle\end{aligned} \tag{3.69}$$

we are able to define an interaction picture state vector via

$$e^{\imath\hat{H}_I t/\hbar}|\psi(0) \tag{3.70}$$

which satisfies the time dependent Schrödinger equation with *just* the interaction Hamiltonian

$$\imath\hbar\frac{\partial\psi_I(t)}{\partial t} = H_I\psi(t). \tag{3.71}$$

and also interaction picture operators

$$\hat{O}_I(t) = e^{\imath\hat{H}_0 t/\hbar}\hat{O}_I e^{-\imath\hat{H}_0 t/\hbar} \tag{3.72}$$

which satisfy equations of the form

$$\frac{d\hat{O}_I}{dt} = \frac{\imath}{\hbar}[\hat{H}_0, \hat{O}_I] + \frac{\partial\hat{O}_S}{\partial t} \tag{3.73}$$

We have cheated just a bit here, we have assumed that we can write

$$\exp(-\imath(\hat{H}_I + \hat{H}_0)t/\hbar) = \exp(-\imath\hat{H}_I t/\hbar)\exp(-\imath\hat{H}_0 t/\hbar) \tag{3.74}$$

We cannot always do this. Recall

$$\begin{aligned}\exp(A + B) &= 1 + (A + B) + (1/2)(A + B)^2 + \cdots \\ &= 1 + (A + B) + (1/2)(A^2 + B^2 + AB + BA) + \cdots\end{aligned} \tag{3.75}$$

On the other hand, we have

$$\begin{aligned}\exp(A)\exp(B) &= (1 + A + (1/2)A^2 + \cdots)(1 + B + (1/2)B^2 + \cdots) \\ &= 1 + (A + B) + (1/2)(A^2 + B^2 + 2AB) + \cdots\end{aligned} \tag{3.76}$$

These two forms are only equivalent if $[A, B] = 0$, which is true for real and complex numbers, but not generally for operators. The implication for the interaction picture is that our above formulae are only correct if we have $[\hat{H}_I, \hat{H}_0] = 0$, which will be true in the cases that interest us most.

3.5 Density matrix

The density operator is defined as

$$\rho = |\psi\rangle\langle\psi| \tag{3.77}$$

where you will recall that $|\psi\rangle$ is the quantum state ket and $\langle\psi|$ is the corresponding bra. For now, the length of the quantum state vector will be some finite N, corresponding perhaps to some internal states of the operator like energy or angular momentum. The expectation value of an operator $\hat{O}$ is given in Dirac notation by

$$\langle\hat{O}\rangle = \langle\psi|\hat{O}|\psi\rangle \tag{3.78}$$

This is just the multiplication of three matrices, a $1xN(\langle\psi|)$, an $N \times N$ (a matrix representation of $\hat{O}$), and finally an $N \times 1$, the ket $|\psi\rangle$. The result is a 1×1 matrix, or scalar, representing the average result of measurements of the observable corresponding to $\hat{O}$ on many identical systems. We can also write this as

$$\begin{aligned}\langle\hat{O}\rangle &= \langle\psi|\hat{O}|\psi\rangle \\ &= \mathrm{Tr}\big(\langle\psi|\hat{O}|\psi\rangle\big) \\ &= \mathrm{Tr}\big(|\psi\rangle\langle\psi|\hat{O}\big) \\ &= \mathrm{Tr}(\rho\hat{O})\end{aligned} \tag{3.79}$$

where we have used the cyclic property of the trace. We can also derive how the density matrix evolves in time, using the Schrödinger equation $i\hbar d|\psi\rangle/dt = \hat{H}|\psi\rangle$,

$$\begin{aligned}\frac{d\rho}{dt} &= \frac{d}{dt}|\psi\rangle\langle\psi| \\ &= \frac{-i}{\hbar}(\hat{H}|\psi\rangle\langle\psi| - |\psi\rangle\langle\psi|\hat{H}) \\ &= \frac{-i}{\hbar}[\hat{H}, \rho]\end{aligned} \tag{3.80}$$

Combining the above two results, we find

$$\begin{aligned}\frac{d\langle\hat{O}\rangle}{dt} &= \frac{dTr(\rho\hat{O})}{dt} \\ &= \mathrm{Tr}\left(\hat{O}\frac{d(\rho)}{dt}\right) \\ &= \frac{-i}{\hbar}\langle[\hat{H}, \hat{O}]\rangle\end{aligned} \tag{3.81}$$

We will have more to say about the density matrix when we deal with dissipation in quantum mechanics.

3.6 Choice of basis and measurement

How do we represent measurement in quantum mechanics? We have described how to get expectation values and standard deviations and the like from the state vector/wave function. The most common measurement made in the laboratory and discussed theoretically is the projective measurement. Let us say we are interested in some physical observable O, which in 1D quantum mechanics is some function $O(x, p)$. If O is physically measurable it must be represented by a Hermitian operator. For the rest of this section, the term measurement is taken to be a projective measurement. The allowed values of a measurement are the eigenvalues of the operator that corresponds to the physical observable. Our meter always gives us a real number, we never get $3 + i$ for an energy measurement, so the fact that we want real values for measurement means we should only consider Hermitian matrices as they are guaranteed to have real eigenvalues. Operators like $\hat{x}$ and $\hat{p}$

are Hermitian. Other operators we shall encounter are not, such as the creation and annihilation operators $\hat{a} = \hat{x} + i\hat{p}$. This means that we cannot measure a physical observable that corresponds to $\hat{a}$.

One other nice property of Hermitian operators is that their eigenvectors form a complete orthogonal set in the case of non-degenerate eigenvalues. We can always find such a set in the case of degenerate eigenvalues too, it just makes more work. This means that we can use the eigenvectors of O to expand our state vector,

$$|\psi\rangle = \sum C_n|o_n\rangle \tag{3.82}$$

where we have defined

$$O|o_n\rangle = o_n|o_n\rangle \tag{3.83}$$

and also $\langle o_n|o_m\rangle = \delta_{n,m}$ We can easily show that $|C_n|^2$ can be interpreted as the probability of finding o_n as a result of measurement of O, and that

$$\langle O\rangle = \sum o_n|C_n|^2 \tag{3.84}$$

We now see that we can use *any* Hermitian operator as a preferred variable, not just position and momentum!

OK, now what is the state of the system after the measurement? Well if the result of the measurement is the value o_n, then the state after the measurement that yielded that value is

$$|\psi_{\text{after}}\rangle = |o_n\rangle \tag{3.85}$$

We can formalize this a bit by introducing a projection operator,

$$P_n = |o_n\rangle\langle o_n| \tag{3.86}$$

Then the state after the measurement is

$$|\psi_{\text{after}}\rangle = \frac{P_n|\psi\rangle}{\sqrt{\langle\psi|P^\dagger P|\psi\rangle}} \tag{3.87}$$

3.7 Executive summary

1. Our quantum system is described by state vector $|\psi\rangle$ in a Hilbert space.
2. The state vector evolves via the Schrödinger equation. The Hermitian nature of the Hamiltonian implies that only unitary evolution is allowed. Any unitary operator has a Hamiltonian operator underlying it, even if it is very complicated. Simple unitaries can have complicated Hamiltonians underlying it.
3. Operators must be represented mathematically in a way that respects the experimental fact that order matters. This is why differential operators and matrices are common ways to represent quantum operators, their operation on a state is not necessarily commutative.

4. Projective measurements of an operator yield results that are eigenvalues of the Hermitian operator that corresponds to the measurable. Recall that Hermitian operators have real eigenvalues (and hence can be related to a single measurement), as well as orthogonal eigenvectors (that can be used as a basis for expanding the wave function).
5. After a projective measurement, the state is in the corresponding eigenvector. Mathematically this is expressed as a projection operator.
6. Other types of measurements, weak measurements and positive operator value measurements are also possible and will be discussed later, in the context of balanced homodyne detection in chapter 7.

Our quantum system is described by state vector $|\psi\rangle$ in a Hilbert space.

3.8 Entanglement

While we will not be examining quantum computation, cryptography, or simulation, we shall be examing *entanglement*, which is a resource for the aforementioned tasks. Entanglement is a property of two quantum systems that are interacting or have interacted. If two systems do not interact, their wave function is a *product* state. Consider two two-level systems, or *qubits*. Each lives in a two-dimensional Hilbert space, and the state of both systems lives in a four-dimensional space. The product space is

$$\mathcal{H}_4 = \mathcal{H}_2^1 \otimes \mathcal{H}_2^2 \tag{3.88}$$

where $\mathcal{H}_2^i$ is a two-dimensional Hilbert space appropriate for the individual systems. A state vector in the larger space, for independent qubits, would be

$$|\Psi_4\rangle = |\psi_2\rangle \otimes |\phi_2\rangle \tag{3.89}$$

Over the last two decades, studies into the foundations of quantum mechanics have shown us that entanglement is an inherently quantum mechanical property, which may be used as a resource for tasks such as factoring a number that is the product of two large prime numbers, search protocols, and quantum teleportation. The measure of entanglement is well defined only in the case of two interacting qubits. Consider a two qubit state

$$|\psi\rangle = C_{00}|11\rangle + C_{01}|01\rangle + C_{10}|10\rangle + C_{11}|00\rangle \tag{3.90}$$

If the system is in a product state

$$|\Psi\rangle = (A_0|0\rangle + A_1|1\rangle)(B_0|0\rangle + B_1|1\rangle) \tag{3.91}$$

Then the coefficients will satisfy

$$\mathcal{E} = C_{00}C_{11} - C_{01}C_{10} = 0 \tag{3.92}$$

If this relation is *not* satisfied then the state is entangled. The concurrence is a measure of entanglement for two qubits and is $\mathcal{C} = \sqrt{2\mathcal{E}}$.

Let us consider some examples. Consider two atoms in an equal superposition of ground and excited states

$$|\psi_i\rangle = \left(\frac{1}{\sqrt{2}}\right)(|g\rangle + |e\rangle) \tag{3.93}$$

the product state of these would be

$$|\Phi\rangle = \left(\frac{1}{2}\right)(|g\rangle + |e\rangle) \otimes (|g\rangle + |e\rangle) \tag{3.94}$$

$$= \frac{1}{2}(|gg\rangle + |ge\rangle + |eg\rangle + |ee\rangle) \tag{3.95}$$

where

$$|eg\rangle = |e\rangle \otimes |g\rangle \tag{3.96}$$

a state where atom 1 is in the excited state and atom 2 is in the ground state.

Now consider a state

$$|+\rangle = \left(\frac{1}{2}\right)(|eg\rangle + |ge\rangle) \tag{3.97}$$

This is a state where the atoms are anticorrelated, in the sense that if one is in the ground state, it is known the other atom is in the excited state and vice versa. So, if I measure the state of atom 1, I have an equal chance of obtaining ground or excited, random. But I know the state of the other atom instantly. If the two atoms are quite far apart this seems like *spooky action at a distance*, but it cannot be used to send information faster than the speed of light as the result of the first measurement is random. It is like a classical particle blowing up into two chunks, momentum is conserved so if I know where one particle is I know where the other chunk is. There is no magic.

This type of state is *not* a product state; in fact, it is maximally entangled, the largest possible concurrence.

What about operators? If we would like an operator that raises atom 1 and lowers atom 2, that would be

$$O = \sigma_+ \otimes \sigma_- \tag{3.98}$$

In matrix form, this would be

$$O = \begin{pmatrix} 0 & 1 \\ 0 & 0 \end{pmatrix} \otimes \begin{pmatrix} 0 & 0 \\ 1 & 0 \end{pmatrix} = \begin{pmatrix} 0 & 0 & 0 & 0 \\ 0 & 0 & 1 & 0 \\ 0 & 0 & 0 & 0 \\ 0 & 0 & 0 & 0 \end{pmatrix}$$

What kind of operator could create an entangled state from $|gg\rangle$? That operator would be

$$O_E = \left(\frac{1}{\sqrt{2}}\right)\begin{pmatrix} 0 & 0 & 0 & 0 \\ 0 & 1 & 0 & 0 \\ 0 & 0 & 1 & 0 \\ 0 & 0 & 0 & 0 \end{pmatrix}$$

It is *not* possible to write this as the outer product of two operators $O_E = O_1 \otimes O_2$. This is a conditional operator, it will put atom 1 in the excited state if the second atom is in the ground state, and vice versa.

References

[1] Ballentine L E 1998 *Quantum Mechanics: A Modern Development* (Singapore: World Scientific)

[2] Beck M 2012 *Quantum Mechanics: Theory and Experiment* (Oxford: Oxford University Press)

[3] Bellac M L and de Forcrand-Millard P 2011 *Quantum Physics* (Cambridge: Cambridge University Press)

[4] Cohen-Tannoudji C, Diu B and Laloe F 1991 *Quantum Mechanics* vol. 1 (New York: Wiley)

[5] Griffiths D J 2017 *Introduction to Quantum Mechanics* (Cambridge: Cambridge University Press)

[6] Sakurai J J and Napolitino J 2006 *Modern Quantum Mechanics* (New York: Springer)

[7] Shankar R 1994 *Principles of Quantum Mechanics* (Berlin: Springer)

IOP Publishing

An Introduction to Quantum Optics

An open systems approach

Perry Rice

Chapter 4

Two-level dynamics

4.1 Two-level atoms

We consider the quantum dynamics of two-level systems now. It is commonly discussed in introductory classes that when shining light (photons/electromagnetic waves) on an atom we can cause the atom to change state, as in figure 4.1.

We could also use molecules, solids, liquids, whatever. This occurs via absorption, in which a photon disappears from the field and the atom goes to a higher energy state. This can occur when the angular frequency of the light, ω, satisfies the Bohr condition $\Delta E = \hbar\omega$, where ΔE is the energy difference between the two states of interest. The atom can also go to a *lower* state (or energy level) by emitting a photon, in a process called stimulated emission. An atom in an excited state will also spontaneously emit to a lower state, which is called spontaneous emission. We will examine these phenomena using the Schrödinger equation, and a simple two-level model. No real atom has only two energy levels, but if the light we shine is near resonance with a transition between two levels we can ignore all other levels as a very good approximation. By 'light', we mean electromagnetic radiation of a well defined frequency, whether it is in the visible, infrared, ultraviolet, or even gamma rays. Our two-level system will be taken to be two levels of an atom, probably a member of the alkali family in practice (cesium, rubidium, and sodium are common in quantum optics experiments), and we will illuminate this atom with monochromatic laser light. Initially, we treat the field as a classical quantity, but in later chapters we more correctly treat it as a quantum mechanical quantity. Many of the results we derive apply to other physical systems, like a nucleus with spin 1/2 in a time varying magnetic field as in nuclear magnetic resonance, but we stick to lasers and atoms as a picture in our heads.

One can more realistically prepare a two-level atom by using optical pumping. Consider the states for $n = 3$ and $n = 4$ as shown below. Driving the system with right circular polarized light will drive the atom up and over one magnetic sublevel. Spontaneous emission can bring you straight down ($\Delta m = 0$) with linearly polarized

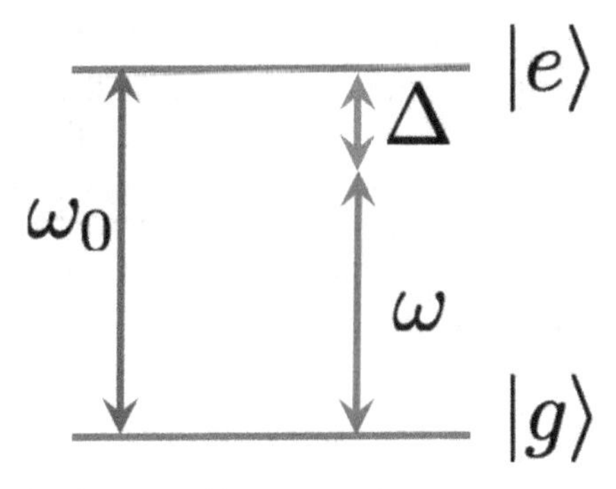

Figure 4.1. Schematic of a two-level system with resonance frequency ω_0, and a driving field at ω, detuned by $\Delta = \omega - \omega_0$.

light, or with right/left circular light ($\Delta m = \pm 1$). Eventually the atom is restricted to the two-state manifold at the far right of the diagram.

The general form for the quantum state of a two-level atom, in the absence of a time dependent interaction potential is

$$|\Psi\rangle = C_e e^{-i\omega_e t}|e\rangle + C_g e^{-i\omega_g t}|g\rangle, \tag{4.1}$$

where the subscripts e and g refer to the excited and ground states, respectively. Here, we have expanded the state vector in the energy eigenbasis, and kept only two terms. The quantities C_e and C_g are probability amplitudes for the excited and ground states. The probability of finding the atom in the excited state (if we actually look!) is $P_e = |C_e|^2$, similarly the probability of finding the atom in the ground state (again, if we actually look!) is $P_g = |C_g|^2$. It is important to recall that in quantum mechanics we know nothing until we make a measurement, and then we only know what we have measured and no more. In the general state equation (4.1), it is *not* as if the atom is sometimes in the ground state and sometimes in the excited state, and hops back and forth. If that were the case, we would merely have statistical physics. In a very real sense, but realizing this is loose language, an atom in state equation (4.1) is simultaneously in the ground *and* excited states. This may seem meaningless, and really it is, as we do not know what state the atom is in until we make a measurement of the energy, then we find that atom in either the ground or excited state, with the probabilities given above. What state is the atom really in in a superposition? In quantum mechanics we do not know anything until we (or someone or something) makes a measurement. As Asher Peres says, 'an unperformed measurement has no outcome'. As we proceed, we will assume that the quantum statics has been done; analytically, numerically, or experimentally—that is, we assume that the eigenstates and eigenvalues of the atom are known to us

$$H_0|e\rangle = \hbar\omega_e|e\rangle \tag{4.2}$$

$$H_0|g\rangle = \hbar\omega_g|g\rangle \tag{4.3}$$

We assume that H_0 is independent of time, and that the resulting eigenstates are stationary. The general superposition is NOT, we recall! But if the system starts out in the ground or excited state, with no admixture from the other, it will sit there in that state for all eternity. In reality, the quantum vacuum *always* provides a time dependent interaction, which is inescapable. For now, we take the eigenstates $|e\rangle$ and $|g\rangle$ as stationary and proceed with the introduction of a time dependent interaction Hamiltonian, due to an external laser field in our later detailed calculations. We take the Hamiltonian to now be

$$H = H_0 + H_I \tag{4.4}$$

We have two avenues open to us. We can take this new Hamiltonian and find the eigenstates and eigenvalues (or eigenenergies). These will be time dependent in general. This leads to the so-called dressed states and quasienergies, and we will examine this path later on. For now, we realize that whatever the state of the system, the states $|e\rangle$ and $|g\rangle$ still form an orthonormal basis, and our state can be expanded in terms of them via

$$|\Psi\rangle = C_e(t)e^{-i\omega_e t}|e\rangle + C_g(t)e^{-i\omega_g t}|g\rangle, \tag{4.5}$$

where now we treat the probability amplitudes C_e and C_g as functions of time; this stems from the fact that the Hamiltonian is dependent on time and will lead to transitions between states. If the system starts in a state such that $C_g(0) = 1$, and $C_e(0) = 0$, i.e. that the atom is in the ground state initially, we will later find that $C_e(t) \neq 0$. The problem will be to find equations for how these probability amplitudes change with time. The probabilities for finding the atom in a given state are also time dependent, for example the excited state probability $P_e(t) = |C_e(t)|^2$. To find the equations the probability amplitudes satisfy, we substitute our Hamiltonian and state vector into the Schrödinger equation. The time derivative of the state vector equation (4.5) is

$$i\hbar\frac{\partial|\Psi\rangle}{\partial t} = i\hbar(\dot{C}_e - i\omega_e C_e)e^{-i\omega_e t}|g\rangle + (\dot{C}_g - i\omega_g C_g)e^{-i\omega_g t}|g\rangle \tag{4.6}$$

The Hamiltonian acting on the state vector is given by

$$\begin{aligned} H|\Psi\rangle &= (H_0 + H_I)|\Psi\rangle \\ &= (\hbar\omega_e + H_I)C_e(t)e^{-i\omega_e t}|e\rangle + (\hbar\omega_g + H_I)C_g(t)e^{-i\omega_g t}|g\rangle \end{aligned} \tag{4.7}$$

where we have used the fact that the states $|e\rangle$ and $|g\rangle$ are eigenstates of the bare Hamiltonian H_0. Using the above results, we obtain

$$\begin{aligned} &i\hbar(\dot{C}_e - i\omega_e C_e)e^{-i\omega_e t}|g\rangle + (\dot{C}_g - i\omega_g C_g)e^{-i\omega_g t}|g\rangle \\ &=(\hbar\omega_e + H_I)C_e(t)e^{-i\omega_e t}|e\rangle + (\hbar\omega_g + H_I)C_g(t)e^{-i\omega_g t}|g\rangle \end{aligned} \tag{4.8}$$

and then using the Bohr condition we find

$$i\hbar\left(\dot{C}_e e^{-i\omega_e t}|e\rangle + \dot{C}_g e^{-i\omega_g t}|g\rangle\right) = \left(C_e(t)e^{-i\omega_e t}H_I|e\rangle + C_g(t)e^{-i\omega_g t}H_I|g\rangle\right) \tag{4.9}$$

To obtain an equation for $\dot{C}_g$, we act on the right of the above equation with the bra $\langle g|$. Using the orthonormality of the excited and ground states, $\langle i|j\rangle = \delta_{ij}$, with i and j either e or g, we have

$$i\hbar \dot{C}_g e^{-i\omega_g t} = \left(C_e(t)e^{-i\omega_e t}\langle g|H_I|e\rangle + C_g(t)e^{-i\omega_g t}\langle g|H_I|g\rangle\right) \tag{4.10}$$

or

$$\dot{C}_g = \frac{-i}{\hbar}\left(C_e(t)e^{-i\omega_0 t}\langle g|H_I|e\rangle + C_g(t)\langle g|H_I|g\rangle\right) \tag{4.11}$$

where we have defined the resonant frequency of the transition as $\omega_0 \equiv \omega_e - \omega_g$. In a similar fashion, we obtain

$$\dot{C}_e = \frac{-i}{\hbar}\left(C_g(t)e^{i\omega_0 t}\langle e|H_I|g\rangle + C_e(t)\langle e|H_I|e\rangle\right) \tag{4.12}$$

We cannot solve these equations until we are told what the interaction Hamiltonian is, and we can compute the matrix elements of the Hamiltonian, $\langle i|H_I|j\rangle$. It is convenient to use a vector representation of the state vector sometimes as

$$|\Psi\rangle = \begin{pmatrix} C_e \\ C_g \end{pmatrix} \tag{4.13}$$

The equations for the probability amplitudes may then be written as

$$\begin{pmatrix} \dot{C}_e \\ \dot{C}_g \end{pmatrix} = -\frac{i}{\hbar}\begin{pmatrix} H_I^{ee} & H_I^{eg}e^{i\omega_0 t} \\ H_I^{eg}e^{-i\omega_0 t} & H_I^{gg} \end{pmatrix}\begin{pmatrix} C_e \\ C_g \end{pmatrix} \tag{4.14}$$

where we have defined $H_I^{ij} \equiv \langle i|H_I|j\rangle$, the matrix elements of the interaction Hamiltonian. For this to be Hermitian, we must have $\langle i|H_I|j\rangle = \langle j|H_I|i\rangle^*$.

4.2 Atom–field interaction in the electric dipole approximation

At this point, we wish to specify an interaction Hamiltonian appropriate for the interaction of an atom with laser light, the electric–dipole interaction

$$H_I = -\vec{\mu} \cdot \vec{E} \tag{4.15}$$

The quantity $\vec{\mu} = -e\vec{x}$ is the dipole moment operator. Recall that for two opposite charges separated by a distance d that the two monopole fields cancel to order $1/r^2$, leaving a residual dipole field proportional to ed/r^3, and $ed = \mu$ is the dipole moment. The total energy of the field is proportional to the square of the total field, $E_{\text{tot}} = E_{\text{app}} + E_{\text{dip}}$. The interaction energy stems from the cross term and goes as $E_{\text{app}}E_{\text{dip}} \propto \mu E_{\text{app}}$. We take $E \equiv E_{\text{app}}$ from now on. The dipole moment is a vector that points from the positive charge to the negative charge and has magnitude ed. In the case of a hydrogen atom, the two charges are the proton and the neutron. For an alkali element, we have a nucleus surrounded by closed orbitals, with a net positive charge, and a single valence electron. These are hydrogen like systems, particularly for circular states where the quantum number m_l is maximized. So, knowing the wave functions of the two states of interest, we could calculate

$$\langle i|H_I|j\rangle = -e\int \psi_i^* \vec{x} \cdot \vec{E}\psi_j dV \tag{4.16}$$

But in any event we will assume that the energies of ground and excited states are known, or really just the difference of the two energies; as well as the interaction matrix elements. This knowledge can be obtained by knowing the wave functions of a tractable system such as hydrogen or a simple harmonic oscillator; numerical solutions of the Schrödinger equation, or measured by spectroscopic means. We will take them to be known and proceed.

The field has contributions from the electric quadrapole and magnetic dipole, as well as electric octopole and magnetic quadrapole terms The electric dipole term is generally the largest, each term in the multipole expansion generally decreases as $1/c$. The key physical assumption is that the electric field does not change much over the size of the atom, and one can approximate $\cos(kx) \simeq 1$. As the size of an atom is in the neighborhood of $0.2 \rightarrow 10.0$ nm, if we consider radiation in the near UV to infrared ($200 \rightarrow 1500$ nm) this condition is easily met. The electric field will not vary much over the location of the electron relative to the (assumed stationary) nucleus. We will take the perturbing field to be of the form $\vec{E} = E_0 \cos(\omega t)\hat{i}$, that is polarized along the x-axis. We shall take it to be propagating along the z-axis, with an amplitude of E_0, and an angular frequency ω that is in general different than the resonant frequency of the atomic transition, ω_0.

The latter will allow us to consider what happens when the laser is not on resonance, but of course it must be nearer this resonance than any others, or else our approximation of a two-level system is in jeopardy. Hence, our interaction Hamiltonian is

$$H_I = -exE_0 \cos(\omega t) \tag{4.17}$$

We will discuss later what happens if the laser is polarized along a different direction, or if one has a collection of atoms whose dipole moments are oriented randomly with respect to the polarization of the laser, as well as circular polarization.

Now let us consider the calculation of the matrix elements of this interaction Hamiltonian. We will assume in what follows that the atomic states of interest have a definite parity, that is $\psi(\vec{x}) = \pm\psi(-\vec{x})$, or rather that $|\psi(\vec{x})|^2$ is an even function. This will be true so long as the probability of the electron to be at a position $\vec{x}$ with respect to the nucleus is the same as the probability for it to be found at $-\vec{x}$; or in more physical terms that the atom has no permanent dipole moment. All of our assumptions to this point can be relaxed but we wish to look at the simplest case first. With this in hand, we first examine a diagonal matrix element.

$$\langle e|H_I|e\rangle = -eE_0 \cos(\omega t)\int \psi_i^* x\psi_j dx = -eE_0\int x|\psi(\vec{x})|^2 dx = 0 \tag{4.18}$$

The same occurs for the other diagonal matrix element, $\langle g|H_I|g\rangle = 0$. There are some systems, particularly molecules, which do have a permanent dipole moment, but we ignore that possibility for now.

Now consider the off-diagonal matrix elements

$$\langle e|H_I|g\rangle = \langle g|H_I|e\rangle^* = -eE_0\cos(\omega t)\int\psi_e^* x\psi_g dx \tag{4.19}$$

This is in general nonzero, and we will assume that the quantity $\langle e|H_I|g\rangle = -eE_0 \cos(\omega t)\int\psi_e^* x\psi_g dx \equiv \mu_{eg}$ is known somehow. Then, our interaction Hamiltonian is given in matrix form by

$$H_I = -i\hbar E_0\cos(\omega t)\begin{pmatrix} 0 & \mu_{eg} \\ \mu_{eg}^* & 0 \end{pmatrix} \tag{4.20}$$

There will be times when $\mu_{eg} = 0$, for example when the states have the same parity. For example, if ψ_e and ψ_g are both odd (or both even), then we will have $\mu_{eg} = 0$. In this case, we are left with $\dot{C}_g = \dot{C}_e = 0$ and there are no transitions. The probabilities $P_e = |C_e(t)|^2$ and $P_g = |C_g(t)|^2$ are stationary. This is the source of selection rules, which tell us which transitions are allowed. For example, the parity of a hydrogen state is given by $P = (-1)^l$, and hence the quantum number l must change in a transition leading to the rule $\Delta l = \pm 1$. A photon is a spin 1 particle and can only carry away one unit of angular momentum. Another selection rule for hydrogen is $\Delta m_l == 0, \pm 1$. These correspond to absorption and emission of circularly polarized photons ($\Delta m_l = \pm 1$), or linearly polarized photons ($\Delta m_l = 0$). Recall that linearly polarized light is a linear combination of right and left circular polarized light. But recall that transitions forbidden for electric dipole interactions can still proceed by higher order terms (for example, electric quadrapole and magnetic dipole), which will result in a smaller amplitude for that component of the spectrum.

If $\Delta m_l = 0$ then for hydrogen states, μ_{eg} is real, as the only complex nature of the matrix element is the ϕ dependence, $\exp(i(m_f - m_i))$,and for convenience we specialize to that case here. The resulting equations are

$$\dot{C}_g = -i\Omega\cos(\omega t)e^{-i\omega_0 t}C_e \tag{4.21}$$

$$\dot{C}_e = -i\Omega\cos(\omega t)e^{i\omega_0 t}C_g \tag{4.22}$$

where we have defined $\Omega = \mu_{eg}E_0/\hbar$, and we will refer to it as the Rabi frequency. Writing the cos in terms of complex exponentials, we find

$$\dot{C}_g = -i\Omega/2(e^{i\Delta t} + e^{-i(\omega_0+\omega)t})C_e \tag{4.23}$$

$$\dot{C}_e = -i\Omega/2(e^{-i\Delta t} + e^{i(\omega_0+\omega)t})C_g \tag{4.24}$$

with $\Delta = \omega - \omega_0$ as the laser detuning from resonance. We will use the rotating wave approximation in the end, which involves dropping the rapidly oscillating terms $e^{i(\omega+\omega_0)t}$; this is known as the rotating wave approximation (RWA). We will explore this approximation more in this chapter, and later when we explore quantum fields. We are then left with

$$\dot{C}_g = \frac{-i\Omega}{2} e^{i\Delta t} C_e \tag{4.25}$$

$$\dot{C}_e = \frac{-i\Omega}{2} e^{-i\Delta t} C_g \tag{4.26}$$

So far, the only approximations made are the dipole approximation and the rotating wave approximation. We will solve these equations using perturbation theory first and then look at the full solutions, on and off resonance. Let us assume that these probability amplitudes have a Taylor series expansion in the parameter Ω,

$$C_e(t) = C_e^{(0)}(t) + C_e^{(1)}(t) + C_e^{(2)}(t) + \cdots \tag{4.27}$$

$$C_e(g) = C_g^{(0)}(t) + C_g^{(1)}(t) + C_g^{(2)}(t) + \cdots \tag{4.28}$$

where (n) indicates the power of Ω for each term. The initial conditions would be

$$C_e^{(0)} = C_e(0) \tag{4.29}$$

$$C_g^{(0)} = C_g(0) \tag{4.30}$$

and for all other values of (n) we have $C_e^{(n)} = C_g^{(n)} = 0$. Plugging this back into equations (4.25) and (4.26), and equating terms in equal powers of Ω, we find

$$\dot{C}_g^{(n)} = \frac{-i\Omega}{2} e^{i\Delta t} C_e^{(n-1)} \tag{4.31}$$

$$\dot{C}_e^{(n)} = \frac{-i\Omega}{2} e^{-i\Delta t} C_g^{(n-1)} \tag{4.32}$$

Another way to develop this formalism is to start with the formal solutions

$$C_e(t) = C_e(0) - \frac{i\Omega}{2} \int_0^t dt' e^{i\Delta t'} C_g(t') \tag{4.33}$$

$$C_g(t) = C_g(0) - \frac{i\Omega}{2} \int_0^t dt' e^{-i\Delta t'} C_e(t') \tag{4.34}$$

and then just iterate it by substituting the solutions on the left-hand side for $C_e(t)$ and $C_g(t)$ back into the formal solutions

$$\begin{aligned} C_e(t) = C_e(0) &- \frac{i\Omega}{2} \int_0^t dt' e^{i\Delta t'} C_g(0) \\ &+ \left(\frac{i\Omega}{2}\right)^2 \int dt' \int dt'' e^{i\Delta(t'-t'')} C_e(t'') \end{aligned} \tag{4.35}$$

$$C_g(t) = C_g(0) - \frac{i\Omega}{2}\int_0^t dt' e^{-i\Delta t'} C_e(0) + \left(\frac{-i\Omega}{2}\right)^2 \int dt' \int dt'' e^{-i\Delta(t'-t'')} C_e(t'') \tag{4.36}$$

One can continue, and in this manner generate a power series for the coefficients, in powers of Ω. For now, we will truncate the series after the first integral, assuming that Ω is small. Recall that this will occur for weak driving fields *or* short interaction times. We also assume that the atom starts in the ground state, so that $C_g(0) = 1$ and $C_e(0) = 0$. This yields

$$C_e(t) = -\frac{i\Omega}{2}\frac{(e^{i\Delta t} - 1)}{i\Delta} \tag{4.37}$$

We then use the following trick

$$e^{ix} - 1 = e^{-ix/2}(e^{-ix/2} - e^{ix/2}) = -2ie^{ix/2}\sin(x/2) \tag{4.38}$$

to write

$$C_e(t) = \frac{i\Omega}{\Delta}e^{i\Delta t/2}\sin(\Delta t/2) \tag{4.39}$$

It will sometimes be convenient to express this in the form

$$C_e(t) = \frac{i\Omega t}{2}e^{i\Delta t/2}\frac{\sin(\Delta t/2)}{\Delta t/2} \tag{4.40}$$

This is convenient to use in the resonant case when $\Delta = 0$, and we find

$$C_e(t) = \frac{i\Omega t}{2}e^{i\Delta t/2} \tag{4.41}$$

recalling that $lim_{x\to 0}\sin(x)/x = 1$. On resonance then, the probability of the atom to be found in the excited state at t, given that it was in the ground state initially when the laser was turned on is

$$P_e(t) = |C_e(t)|^2 = \frac{\Omega^2 t^2}{4} \tag{4.42}$$

Obviously, if we let t get larger and larger, the probability for being in the excited state exceeds one! Well all this means is that we can no longer ignore the higher order terms in the perturbation series above, which goes as powers of Ωt. In fact, the above result defines what we mean by a 'short' time or 'weak' field, it is when

$$t \ll 2/\Omega \tag{4.43}$$

When we are off resonance, the result is

$$P_e(t) = |C_e(t)|^2 = \frac{\Omega^2\sin^2(\Delta t/2)}{\Delta^2} \tag{4.44}$$

Plots of this probability for various detunings are shown in figure 4.2.

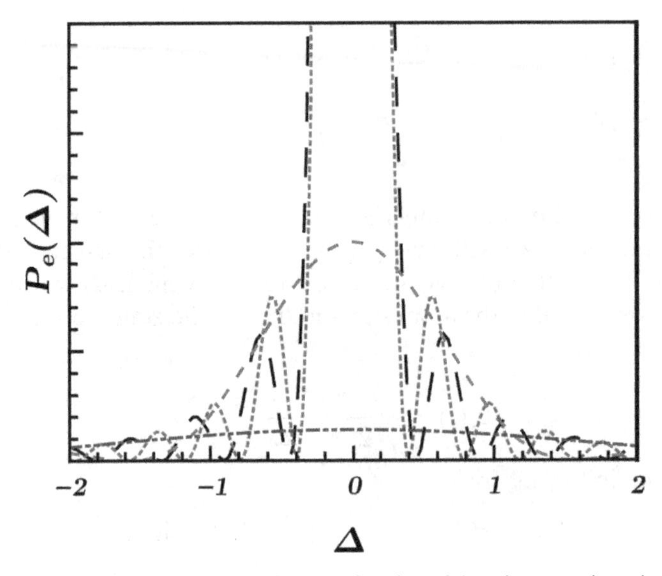

Figure 4.2. Excited state population as a function of detuning at various times.

Perhaps the surprising result here is that even if one is detuned from resonance, at a given time it is possible to find the atom in the excited state! Well that seems to be at odds with energy conservation; we are putting in photons with energy $\hbar\omega$, which is larger or smaller than the energy difference between the two levels $\hbar\omega_0$. This is a problem for you to figure out! Think about how that atom knows what frequency is illuminating it, and when!

If we had kept the non-RWA terms in equation (4.24) that oscillated at $\omega + \omega_0$, the resulting integrals would have yielded terms like

$$\frac{e^{i(\omega+\omega_0)t} - 1}{i(\omega + \omega_0)} \tag{4.45}$$

The numerator of this term would oscillate between 0 and 2 in amplitude while the denominator is of order 10^{15} for visible light. Obviously, in perturbation theory, these terms will have little effect. Later when we quantize the field we will find they make no contributions to transition rates in the dipole approximation. Let us now go beyond perturbation theory, beginning on resonance. The relevant equations are

$$\dot{C}_g = \frac{-i\Omega}{2} C_e \tag{4.46}$$

$$\dot{C}_e = \frac{-i\Omega}{2} C_g \tag{4.47}$$

We take the derivative of the first equation which yields

$$\ddot{C}_e = \frac{-i\Omega}{2} \dot{C}_g = \frac{-\Omega^2}{4} C_e \tag{4.48}$$

This has the solution

$$C_e(t) = A\cos(\Omega t/2) + B\sin(\Omega t/2) \tag{4.49}$$

Knowing C_e, we can use the differential equation for $\dot{C}_e$ to find C_g,

$$C_g(t) = iA\sin(\Omega t/2) - iB\cos(\Omega t/2) \tag{4.50}$$

We have the normalization condition $P_e + P_g = |A|^2 + |B|^2$, where A and B are generally complex and found from initial conditions.

With the atom initially in the ground state, we have $C_e(0) = A = 0$, and $C_g(0) = 1 = iB$, or $B = -i$. Thus, we have

$$C_e(t) = i\sin(\Omega t/2) \tag{4.51}$$

$$C_g(t) = \cos(\Omega t/2) \tag{4.52}$$

The associated probabilities are

$$P_e(t) = \sin^2(\Omega t/2) = \frac{1}{2}(1 - \cos(\Omega t)) \tag{4.53}$$

$$P_g(t) = \cos^2(\Omega t/2) = \frac{1}{2}(1 + \cos(\Omega t)) \tag{4.54}$$

This is shown in figure 4.3.

Again we see that the sum of probabilities is unity at all times, which must occur due to the unitary evolution of the quantum state. The probabilities oscillate at the frequency Ω, known as the Rabi frequency. Notice that this exact result reduces to

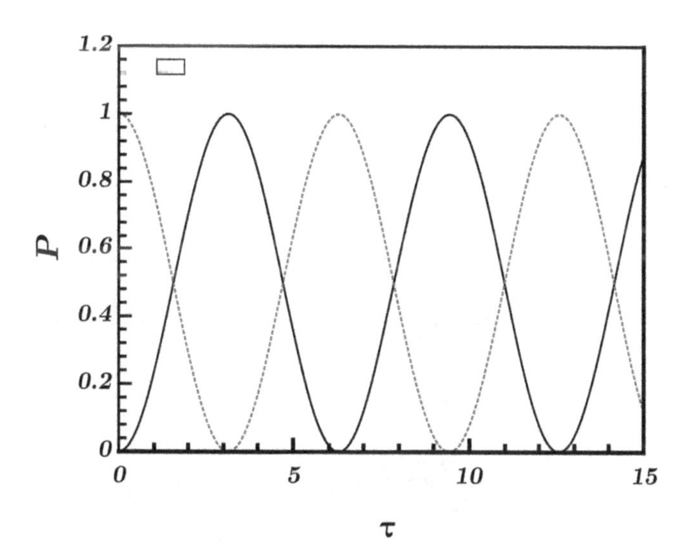

Figure 4.3. Excited and ground state populations as functions of time when the system is driven on resonance.

our perturbation result for $\Omega t \ll 1$, where we have previously determined that perturbation theory would hold.

Notice that turning the laser off at a time $\Omega T = \pi$, the atom is in the excited state, with no admixture from the ground state. This is referred to as a π pulse. Another special case is when the laser is turned off at $T = \pi/2\Omega$ or $\Omega T = \pi/2$. Here, the atom is left in the superposition state

$$|\Psi\rangle = \frac{1}{\sqrt{2}}(i|e\rangle + |g\rangle) \tag{4.55}$$

The atom is in a state where the atom is equally likely to be found in the ground or excited states if an energy measurement is made, but it is neither in the ground or excited state *unless* such a measurement is made. A 2π pulse would return the atom to the ground state. Such pulses can be useful for communication applications as no energy is deposited in the atoms of, say, the fiber. This can only occur in realistic systems for certain pulse shapes, which result in the phenomena of solitons and self-induced transparency. If the laser is not turned on and off abruptly (is it ever????) the relevant parameter is the pulse 'area' $A = \int_0^T \Omega(t)dt$.

What about off resonance? Well our trick of differentiating one of the equations and substituting that into the other will still work, but it is a little more complicated, resulting in

$$\ddot{C}_e + i\Delta\dot{C}_e + \frac{\Omega^2}{4}C_e = 0 \tag{4.56}$$

$$\ddot{C}_g - i\Delta\dot{C}_e + \frac{\Omega^2}{4}C_e = 0 \tag{4.57}$$

These equations can be solved using the ansatz $C_e(t) = e^{i\lambda t}$. The first equation then yields

$$\left[-\lambda^2 - \lambda\Delta + \frac{\Omega^2}{4}\right] = 0 \tag{4.58}$$

The two roots are

$$\lambda_{1,2} = -\frac{1}{2}(\Delta \pm \Omega'^2) \tag{4.59}$$

where the generalized Rabi frequency is $\Omega' = \sqrt{\Omega^2 + \Delta^2}$. The general form of the solutions is

$$C_e(t) = Ae^{i\lambda_1 t} + Be^{i\lambda_2 t} \tag{4.60}$$

$$C_g(t) = \frac{-2i}{\Omega}e^{i\Delta t/2}[\lambda_1 Ae^{i\lambda_1 t} + \lambda_2 Be^{i\lambda_2 t}] \tag{4.61}$$

With the atom initially in the ground state, we have

$$C_e(t) = -i\frac{\Omega}{\Omega'}\sin(\Omega' t/2)e^{i\Delta t/2} \tag{4.62}$$

$$C_g(t) = \cos(\Omega' t/2) - i\frac{\Delta}{\Omega'}\sin(\Omega' t/2)e^{-i\Delta t/2} \tag{4.63}$$

A plot of the excited state probability is shown in figure 4.4. As we increase the detuning, we find that the maximum excited state probability decreases, and the oscillations become faster, oscillating at the generalized Rabi frequency.

Notice that the probability oscillates at a larger frequency $\Omega' \gg \Omega$ than on resonance, and also note that the atom never reaches the excited state.

Notice that we have never explicitly used the fact that the excited state had a greater energy than the ground state, so our results also work for stimulated emission as well. Actually, there are some phase changes for emission, but the probabilities remain the same. Mathematically, one would make the substitution $\omega_0 = -|\omega_0|$, and the terms dropped in the RWA would be kept. But the terms we kept would then involve sums of frequencies and would be dropped.

4.3 Introduction to dressed states

Until now we have used the excited and ground states as basis states to expand a general quantum state of the two-level atom; let us now consider a different tack and talk about the eigenstates and eigenvalues of the Hamiltonian $H = H_0 + H_I$. These eigenstates are time dependent as the Hamiltonian is, they are *not* stationary. However, if we make a time-dependent transformation to our probability equations, we can find some useful results. This transformation is

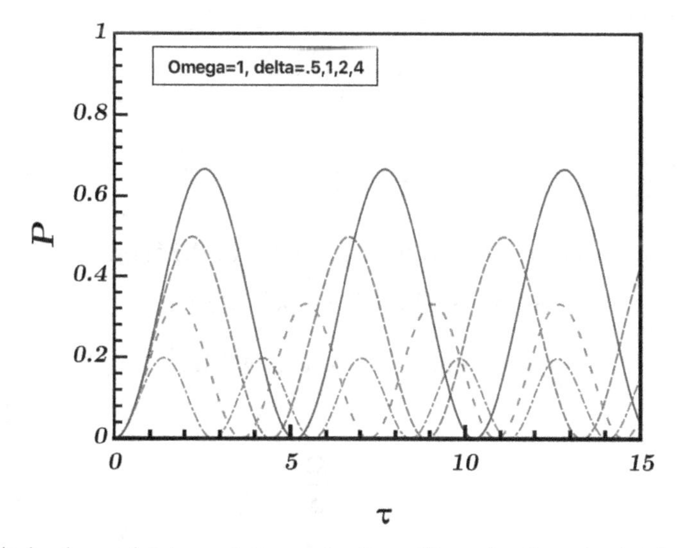

Figure 4.4. Excited and ground state populations as functions of time when the system is driven off resonance.

$$D_g(t) = C_g(t) \tag{4.64}$$

$$D_e(t) - C_e(t)e^{-i\Delta t} \tag{4.65}$$

The new probability amplitudes satisfy the equations

$$\dot{D}_g = \frac{-i\Omega}{2} D_e \tag{4.66}$$

$$\dot{D}_e = \frac{-i\Omega}{2} D_g - i\Delta D_e \tag{4.67}$$

These equations can be written in the form

$$\begin{pmatrix} \dot{D}_e \\ \dot{D}_g \end{pmatrix} = -\frac{i}{\hbar}\begin{pmatrix} -\hbar\Delta & \hbar\Omega/2 \\ \hbar\Omega/2 & 0 \end{pmatrix}\begin{pmatrix} D_e \\ D_g \end{pmatrix} \tag{4.68}$$

In this new frame, our new Hamiltonian has no time dependence, the eigenvalues or quasi-energies are

$$E_{1,2} = \frac{\hbar}{2}(-\Delta \pm \Omega') \tag{4.69}$$

The new (unnormalized) eigenstates are

$$|+\rangle = \begin{pmatrix} \Omega' - \Delta \\ \Omega \end{pmatrix} \tag{4.70}$$

$$|-\rangle = \begin{pmatrix} \Omega' + \Delta \\ -\Omega \end{pmatrix} \tag{4.71}$$

On resonance we have

$$|+\rangle = \frac{1}{\sqrt{2}}\begin{pmatrix} 1 \\ 1 \end{pmatrix} \tag{4.72}$$

$$|-\rangle = \frac{1}{\sqrt{2}}\begin{pmatrix} 1 \\ -1 \end{pmatrix} \tag{4.73}$$

It is often convenient to express the eigenstates in the form

$$|+\rangle = \cos(\Theta/2)|e\rangle + \sin(\Theta/2)|g\rangle \tag{4.74}$$

$$|-\rangle = -\sin(\Theta/2)|e\rangle + \cos(\Theta/2)|g\rangle \tag{4.75}$$

with

$$\tan(\Theta) = \frac{\Omega}{\Delta} \tag{4.76}$$

Again, on resonance, we have $\Theta = \pi$, with $\Theta/2 = \pi/2$, with

$$|+\rangle = \frac{1}{\sqrt{2}}(|e\rangle + |g\rangle) \tag{4.77}$$

$$|-\rangle = \frac{1}{\sqrt{2}}(|g\rangle - |e\rangle) \tag{4.78}$$

The two states effectively split into two, separated by 2Ω. The details of this doubling are a hallmark of Floquet states, of which the dressed states are an example. It is obvious that we will get spectral peaks at ω_0 as well as sidebands at $\omega_0 \pm \Omega$.

Very far off resonance, the energies are

$$E_1 = E_g + \frac{\hbar\Omega^2}{4\Delta} \tag{4.79}$$

$$E_2 = E_e - \frac{\hbar\Omega^2}{4\Delta} \tag{4.80}$$

The energy shifts quadratic in the driving field are known as 'light shifts', or AC Stark shifts as shown in figure 4.5. In some situations for cooling and trapping of atoms, the field (and hence Ω and the light shifts) is position dependent, leading to an energy gradient, or force! More later.

4.4 Perturbation theory and rate equations

We now return to perturbation theory for an examination of what happens when we drive the atom with many colors of light at once, instead of monochromatic excitation. Recall that for one color of light we had

$$P_e(t) = \frac{\Omega^2 t^2}{4} \frac{\sin^2(\Delta t/2)}{(\Delta t/2)^2} \tag{4.81}$$

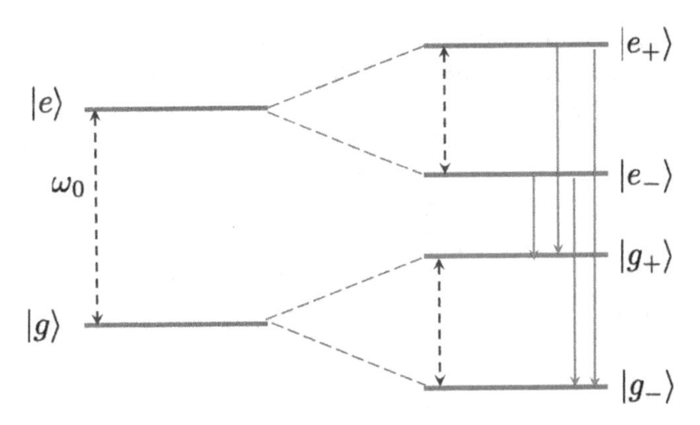

Figure 4.5. Dressed states.

If we now shine light of many colors on the atom, and this light is completely incoherent, we may sum probabilities instead of summing amplitudes and then squaring to get a probability. We also use $E_0^2 = 2U(\Omega)/\varepsilon_0$ where $U(\omega) = |E_0|^2/\varepsilon_0$ is the energy density of the field at frequency ω. For the incoherent driving field, we then have

$$P_e(t) = \frac{|\mu_{eg}|^2 t^2}{4\varepsilon_0 \hbar^2} \frac{2}{t} \int_{-\infty}^{\infty} U(\Delta) \frac{\sin^2(\Delta t/2)}{(\Delta t/2)^2} d\left(\frac{\Delta t}{2}\right) \tag{4.82}$$

The definite integral has no time dependence, so the probability of a transition goes linearly with time. We have integrated over all detunings and extended the lower limit of integration to $-\infty$. This is a very good approximation as $U(\Delta)$ multiplies a function that dies off to zero at large detunings, a $\mathrm{sinc}(x) = \sin(x)/x$ function. We next assume that the energy density of the driving field is slowly varying over the sinc function in the integral and approximate $U(\Delta) \approx U(0)$ and pull it out of the integral. The definite integral can then be evaluated, leading to a probability of being in the excited state proportional to t instead of t^2. The final result is

$$P_e(t) = \frac{\pi |\mu_{eg}|^2}{2\varepsilon_0 \hbar^2} U(\omega_0) t \equiv BU(\omega_0) t \tag{4.83}$$

where we have defined the Einstein B coefficient $B = \pi|\mu_{eg}|^2/2\varepsilon_0\hbar^2$. If we average over the orientation of the atom's dipole moment, we gain an additional factor of 1/3, due to a spatial averaging over the $\cos(\theta_{E,\mu})$ resulting from the dot product of the applied field and the dipole moment, resulting in $B = \pi|\mu_{eg}|^2/6\varepsilon_0\hbar^2$. This result is a special case of Fermi's golden rule, where an incoherent excitation leads to a transition probability that is linear in time and proportional to the square of a matrix element of the interaction Hamiltonian. We can define a transition rate $R_{\mathrm{abs}} = dP_e/dt = BU(\omega_0)$, which is a constant. We also have (assuming no degeneracy of ground and excited states as we have) $R_{\mathrm{abs}} = R_{\mathrm{stimem}}$. What if the light source is monochromatic, but the atoms in our sample have different resonant frequencies, due to Doppler or pressure broadening? Then our result still holds. We only need to have a uniform range of detunings to integrate over, hence we can use rate equations for treatments of transitions that are subject to broadening mechanisms when pumped with monochromatic light.

Let us now consider many atoms being excited incoherently. The ground and excited state populations change via three processes: absorption, stimulated emission, and spontaneous emission. The latter process occurs when there is no applied field. In all of our previous work, we have found transitions, but *only* when we have an applied field. We will follow Einstein's thinking and look at rate equations for the populations in the two states.

$$\dot{N}_e = BU(\omega_0)(N_g - N_e) - AN_e \tag{4.84}$$

$$\dot{N}_g = BU(\omega_0)(N_e - N_g) + AN_e \tag{4.85}$$

Notice that the total number of atoms does not change, $dN/dt = dN_e/dt + dN_g/dt = 0$. We will follow Einstein's thinking and consider the case of thermal equilibrium. In that case, we know that the energy density of the field is given by the Planck distribution

$$U(\omega_0) = \frac{\hbar\omega^3}{c^3\pi^2}\frac{1}{e^{\hbar\omega/kT} - 1} \tag{4.86}$$

Also, the time derivatives are zero as we are in equilibrium. Further, we know that the relative number of atoms in the ground and excited states is given by the Boltzmann distribution (assuming that the mean free path is large compared to the deBroglie wavelength of the atoms)

$$N_e/N_g = e^{-\hbar\omega_0/kT} \tag{4.87}$$

All that is left is to find A! Rearranging, we find

$$A = \frac{\hbar\omega^3 B}{\pi^2 c^3} = \frac{\hbar\omega^3\mu_{eg}^2}{6\pi\varepsilon_0\hbar^2 c^3} \tag{4.88}$$

It is interesting to see how leaving out the stimulated emission term affects the results. At the turn of the century, only absorption and spontaneous emission were observed in the laboratory. We would then have from our rate equation analysis

$$BU(\omega_0)N_g - AN_e = 0 \tag{4.89}$$

This leads to

$$U(\omega_0) = (A/B)e^{-\hbar\omega_0/kT} \tag{4.90}$$

This is certainly not the Planck distribution at all, but the distribution derived by Wien based on Maxwell's equations. Einstein's purpose was to learn more about the Planck distribution, and he found that by adding the stimulated emission term to the two known processes he could derive the Planck distribution from very basic principles. This was in 1905, and no one had calculated B yet! It was 20 years before the Schrödinger equation was discovered. Einstein showed that to explain the data in 1905, the process of stimulated emission must exist; he is the father of the laser. We can derive A today from a treatment that treats the field quantum mechanically, done first by Dirac. But where in Einstein's treatment is the field treated quantum mechanically? In using the Planck distribution, which was where $\hbar$ first appeared.

What if we ignore the spontaneous emission term? After all, we have only put it in by hand. Then we would have

$$BU(\omega_0)(N_g - N_e) = 0 \tag{4.91}$$

which leads to $N_g = N_e$, and this means that we would have gotten the atomic statistics wrong. One needs all three processes to get everything right.

4.5 Pauli operators and the Bloch sphere representation

It is oftentimes useful to consider a real vector space instead of a complex one. In the case of our two-level atom, we have a two-dimensional Hilbert space, and a wave function

$$|\psi\rangle = C_e|e\rangle + C_g|g\rangle \tag{4.92}$$

where C_e and C_g are complex numbers specified by four real numbers (either real and imaginary parts, or magnitudes and phases). Thus our two-dimensional Hilbert space can be represented as a four-dimensional real vector space. We can write our wave function as

$$\begin{aligned} |\psi\rangle &= |C_e|e^{i\phi_e}|e\rangle + |C_g|e^{i\phi_g}|g\rangle \\ &= e^{i\phi_e}\left(|C_e||e\rangle + |C_g|e^{i(\phi_g-\phi_e)}|g\rangle\right) \\ &= |C_e||e\rangle + |C_g|e^{i(\phi_g-\phi_e)}|g\rangle \end{aligned} \tag{4.93}$$

where in the last step we have ignored the overall phase $e^{i\phi_e}$. So, given the global phase invariance used here, we now have only three real numbers to describe our quantum state. We can also use the relation $|C_e|^2 + |C_g|^2 = 1$ to reduce our description of the two-level system to two real variables

$$\begin{aligned} |\psi\rangle &= |C_e||e\rangle + \sqrt{1-|C_e|^2}\, e^{i(\phi_g-\phi_e)}|g\rangle \\ &= \sin(\theta/2)|e\rangle + \cos(\theta/2)e^{i\phi/2}|g\rangle \end{aligned} \tag{4.94}$$

where we have defined $C_e = \sin(\theta/2)$ and $\phi_g - \phi_e = \phi/2$. The factors of two stem from the fact that a π rotation in the Bloch sphere space is a $\pi/2$ rotation in the two-dimensional complex space that our states and operators live in. We treat θ and ϕ as a latitude and longitude, and we can represent any quantum state of a two-level system on the Bloch sphere, as shown below.

As our quantum state is described by a complex vector of length two, all operators that act on the state are unitary 2×2 matrices,

$$U = \begin{pmatrix} a & b \\ c & d \end{pmatrix} \tag{4.95}$$

with $ad - bc = \det(U) = 1$. Just as a vector can be expanded in terms of a set of linearly independent basis vectors, $\vec{V} = V_x\hat{i} + V_y\hat{j} + V_z\hat{k}$, one can expand a matrix in terms of linearly independent basis matrices,

$$\begin{aligned} U = a&\begin{pmatrix} 1 & 0 \\ 0 & 0 \end{pmatrix} + b\begin{pmatrix} 0 & 1 \\ 0 & 0 \end{pmatrix} \\ c&\begin{pmatrix} 0 & 0 \\ 1 & 0 \end{pmatrix} + d\begin{pmatrix} 0 & 0 \\ 0 & 1 \end{pmatrix} \end{aligned} \tag{4.96}$$

OK, this is not so impressive and may seem a waste of time. However, besides the obvious choice of basis matrices above, it is useful to consider the Pauli operators

$$\sigma_x = \begin{pmatrix} 0 & 1 \\ 1 & 0 \end{pmatrix} \quad (4.97)$$

$$\sigma_y = \begin{pmatrix} 0 & -i \\ i & 0 \end{pmatrix} \quad (4.98)$$

$$\sigma_z = \begin{pmatrix} 1 & 0 \\ 0 & -1 \end{pmatrix} \quad (4.99)$$

These operators, plus the identity operator I, can be used as a complete set of operators for a two-level system, as any operator can be written as

$$O = AI + B\sigma_x + C\sigma_y + D\sigma_z \quad (4.100)$$

The Pauli operators satisfy the following properties:

$$Tr(\sigma_i) = 0 \quad (4.101)$$

$$\sigma_i\sigma_j = -\sigma_j\sigma_i \quad (4.102)$$

$$(\sigma_i)^2 = I \quad (4.103)$$

$$[\sigma_i, \sigma_j] = 2i\varepsilon_{ijk}\sigma_k \quad (4.104)$$

$$(\vec{\sigma} \cdot \hat{n})^2 = 1 \quad (4.105)$$

$$[\sigma_i, \sigma_j]_+ = 2\delta_{ij} \quad (4.106)$$

Here, $\vec{\sigma}$ is a vector whose three components are the three Pauli operators, and $\hat{n}$ is any unit vector in three dimensions. Also, ε_{ijk} is the Levi-Civita symbol, a fully antisymmetric tensor of rank 2. More simply, it is zero if any index repeats, +1 if the indices are in any cyclic order of 1, 2, 3, and −1 if 1, 2, 3 appear in a non-cyclic order, such as 321. These operators obey the same algebra, known as the Clifford algebra, that angular momentum operators satisfy and in fact are the appropriate operators to deal with a spin 1/2 system. This is not surprising as a spin 1/2 system has two states, spin up and spin down. So the math for the spin 1/2 system is the same as for our two-level atom, as they are both two state systems. There is another relation that is very useful for manipulating Pauli operators,

$$exp(i\vec{\theta} \cdot \sigma/2) = \cos(\theta/2)I + i(\hat{\theta} \cdot \vec{\sigma}/2)\sin(\theta/2) \quad (4.107)$$

which can be verified by a power series expansion of both sides. Here, $\vec{\theta}$ is a vector with a magnitude θ, and some direction given by the unit vector $\hat{\theta}$. Recall that any state of our two-level system can be represented in two ways: one as a vector in a two-dimensional complex vector space, or as we have seen, in a three-dimensional real space, the Bloch sphere. In either case, any operator that acts on a state to generate a new state can be viewed as a rotation of the vector, so we need to be able to manipulate rotation operators of the form of equation (4.107). Let us focus on the Bloch sphere representation for now, here a position on the Bloch sphere can be determined by the latitude (θ) and longitude (ϕ). A unit vector that points at this point is

$$\hat{n} = \begin{pmatrix} \sin(\theta/2)\cos(\phi/2) \\ \sin(\theta/2)\sin(\phi/2) \\ \cos(\theta/2) \end{pmatrix} \tag{4.108}$$

We use half angles to be consistent with our earlier work. Two other operators that are variations of the Pauli operators that will be useful are

$$\sigma_{\pm} = \frac{1}{2}(\sigma_x \pm \sigma_y) \tag{4.109}$$

with matrix representations

$$\sigma_+ = \begin{pmatrix} 0 & 1 \\ 0 & 0 \end{pmatrix} \tag{4.110}$$

$$\sigma_- = \begin{pmatrix} 0 & 0 \\ 1 & 0 \end{pmatrix} \tag{4.111}$$

The action of these operators on the ground and excited states

$$\begin{aligned} |e\rangle &= \begin{pmatrix} 1 \\ 0 \end{pmatrix} \\ |g\rangle &= \begin{pmatrix} 0 \\ 1 \end{pmatrix} \end{aligned} \tag{4.112}$$

are

$$\begin{aligned} \sigma_+|e\rangle &= 0 \\ \sigma_+|g\rangle &= |e\rangle \\ \sigma_-|e\rangle &= |g\rangle \\ \sigma_-|g\rangle &= 0 \end{aligned} \tag{4.113}$$

We refer to σ_+ as a raising operator, as it takes the atom from the ground state to the excited state. It will obviously play a role in describing the absorption of a photon. Then σ_- is referred to as a lowering operator and will be useful in describing emission.

The matrix representation of the Hamiltonian for a two-level atom, in the absence of a driving field is

$$H_0 = \begin{pmatrix} E_e & 0 \\ 0 & E_g \end{pmatrix} = \frac{(E_e + E_g)}{2} I + \frac{(E_e - E_g)}{2} \sigma_z \tag{4.114}$$

The identity term can be eliminated by redefining the zero of energy to be midway between the excited and ground states, which leaves us with

$$H_0 = \frac{\hbar\omega_0}{2}\sigma_z \tag{4.115}$$

where again $E_e - E_g = \hbar\omega_0$. Starting with the initial state

$$|\psi(0)\rangle = \begin{pmatrix} C_e(0) \\ C_g(0) \end{pmatrix} \tag{4.116}$$

we can solve for the time dependence via the Schrödinger equation

$$|\psi(t)\rangle = e^{-iH_0t/\hbar}|\psi(0)\rangle = e^{-i\omega_0\sigma_z t/2}|\psi(0)\rangle \tag{4.117}$$

The exponentiation of an operator has meaning via its Taylor series expansion, and as σ_z is a diagonal representation, we find that

$$e^{-iH_0t/\hbar} = \begin{pmatrix} e^{-i\omega t/2} & 0 \\ 0 & e^{i\omega t/2} \end{pmatrix} \tag{4.118}$$

which leads to

$$|\psi(t)\rangle = \begin{pmatrix} C_e(0)e^{-\iota\omega t/2} \\ C_g(0)e^{\iota\omega t/2} \end{pmatrix} \tag{4.119}$$

which is the form we used earlier.

The electric dipole Hamiltonian, $H_I = -\vec{\mu} \cdot \vec{E}$ has the matrix representation

$$H_I = -\mu_{eg}^z E_z \begin{pmatrix} 0 & 1 \\ 1 & 0 \end{pmatrix} = -\mu_{eg}^z E_z\sigma_x = -\mu_{eg}^z E_z(\sigma_+ + \sigma_-) \tag{4.120}$$

where we have taken the driving electric field to be polarized along the z direction. We can see that the raising and lowering operators are involved, dealing with absorption and emission terms as we alluded to earlier. We recall that in the Heisenberg picture, the time evolution is carried by the operators,

$$\begin{aligned} \frac{d\hat{O}_H}{dt} = & \frac{\iota}{\hbar}\hat{H}e^{\iota\hat{H}t/\hbar}\hat{O}_H e^{-\iota\hat{H}t/\hbar} \\ & + e^{\iota\hat{H}t/\hbar}\frac{\partial\hat{O}_H}{\partial t}e^{-\iota\hat{H}t/\hbar} \\ & - \frac{\iota}{\hbar}e^{\iota\hat{H}t/\hbar}\hat{O}_H e^{-\iota\hat{H}t/\hbar} \end{aligned} \tag{4.121}$$

In our case, we are concerned with the three Pauli operators, for example

$$\begin{aligned}\frac{d\sigma_x}{dt} &= [H_0 + H_I, \sigma_+] \\ &= \frac{\imath\omega_0}{2}[\sigma_z, \sigma_x] + \frac{\imath\Omega_0}{2}e^{i\omega t}[\sigma_x, \sigma_x] \\ &= -\omega_0\sigma_y\end{aligned} \tag{4.122}$$

Similarly, we find that

$$\frac{d\sigma_y}{dt} = \omega_0\sigma_x + 2\Omega_0 \cos \omega t \sigma_z \tag{4.123}$$

$$\frac{d\sigma_z}{dt} = -2\Omega_0 \cos \omega t \sigma_y \tag{4.124}$$

In terms of raising and lowering operators, we find

$$\frac{d\sigma_+}{dt} = i\omega_0\sigma_+ + i\Omega_0 \cos(\omega t)\sigma_z \tag{4.125}$$

$$\frac{d\sigma_-}{dt} = -i\omega_0\sigma_- + i\Omega_0 \cos(\omega t)\sigma_z \tag{4.126}$$

$$\frac{d\sigma_z}{dt} = 2i\Omega_0 \cos \omega t(\sigma_+ - \sigma_-) \tag{4.127}$$

It is often convenient to go to a rotating frame, via

$$\sigma_+ = \Sigma_+ e^{i\omega t} \tag{4.128}$$

$$\sigma_- = \Sigma_- e^{-i\omega t} \tag{4.129}$$

$$\sigma_z = \Sigma_z \tag{4.130}$$

This leads to equations for the slowly varying Σ operators

$$\frac{d\Sigma_+}{dt} = i\Delta\Sigma_+ + i\frac{\Omega_0}{2}(1 + e^{-2i\omega t})\Sigma_z \tag{4.131}$$

$$\frac{d\Sigma_-}{dt} = -i\Delta\Sigma_+ - i\frac{\Omega_0}{2}(1 + e^{-\omega t})\Sigma_z \tag{4.132}$$

$$\frac{d\Sigma_z}{dt} = +i\Omega_0(\Sigma_+ - \Sigma_-) \tag{4.133}$$

where we have $\Delta = \omega_0 - \omega$ as before. If we drop rapidly oscillating terms (2ω), which is the rotating wave approximation, we find

$$\frac{d\Sigma_+}{dt} = i\Delta\Sigma_+ + i\frac{\Omega_0}{2}\Sigma_z \tag{4.134}$$

$$\frac{d\Sigma_-}{dt} = -i\Delta\Sigma_+ - i\frac{\Omega_0}{2}\Sigma_z \tag{4.135}$$

$$\frac{d\Sigma_z}{dt} = +i\Omega_0(\Sigma_+ - \Sigma_-) \tag{4.136}$$

We could have invoked the rotating wave approximation earlier by using the interaction Hamiltonian

$$H_I = -\frac{\hbar\Omega_0}{2}(\sigma_+ e^{-i\omega t} + \sigma_- e^{i\omega t}) \tag{4.137}$$

It is worth noting that an equivalent Hamiltonian would be

$$H_I = i\frac{\hbar\Omega_0}{2}(\sigma_+ e^{-i\omega t} - \sigma_- e^{i\omega t}) \tag{4.138}$$

This basically would involve a phase shift of the driving field, or induced dipole moment, of $\pi/2$. This would lead to the following equations:

$$\frac{d\Sigma_+}{dt} = i\Delta\Sigma_+ + \frac{\Omega_0}{2}\Sigma_z \tag{4.139}$$

$$\frac{d\Sigma_-}{dt} = -i\Delta\Sigma_+ - + \frac{\Omega_0}{2}\Sigma_z \tag{4.140}$$

$$\frac{d\Sigma_z}{dt} = -\Omega_0(\Sigma_+ - \Sigma_-) \tag{4.141}$$

The i in the new Hamiltonian is canceled by the one in the Heisenberg equation of motion, so on resonance we are dealing only with real quantities. We will utilize the earlier version with the i's for now.

Is there a relation between the Heisenberg equations above, and the $\dot{C}$ equations of the Schrödinger picture? And what is the physical interpretation of these equations? To make these connections we examine the expectation values of the Pauli operators. For example

$$\begin{aligned}\langle\sigma_z\rangle &= \langle\psi|\sigma_z|\psi\rangle \\ &= (C_e^* C_g^*)\begin{pmatrix}1 & 0 \\ 0 & -1\end{pmatrix}\begin{pmatrix}C_e \\ C_g\end{pmatrix} \\ &= |C_e|^2 - |C_g|^2\end{aligned} \tag{4.142}$$

We see that the expectation value of Σ_z is the population inversion of the atom. What about the raising and lowering operators? Consider the induced dipole operator

$$\begin{aligned}\mu_{eg}\begin{pmatrix}0 & 1\\ 1 & 0\end{pmatrix} &= \mu_{eg}\sigma_x\\ &= \mu_{eg}(\sigma_+ + \sigma_-)\\ &= \mu_{eg}(\Sigma_+ e^{i\omega t} + \Sigma_- e^{-i\omega t})\\ &= \mu_{eg}[(\Sigma_+ + \Sigma_-)\cos\omega t + i(\Sigma_+ - \Sigma_-)\sin\omega t]\end{aligned} \tag{4.143}$$

The expectation value of σ_+ is given by

$$\begin{aligned}\langle\sigma_z\rangle &= \langle\psi|\sigma_z|\psi\rangle\\ &= (C_e^* C_g^*)\begin{pmatrix}0 & 1\\ 0 & 0\end{pmatrix}\begin{pmatrix}C_e\\ C_g\end{pmatrix}\\ &= C_e^* C_g\end{aligned} \tag{4.144}$$

This shows us that basically the expectation values of $\sigma_\pm$ are the average induced dipole moment of the atom. Let us refer back to the Bloch sphere picture for a bit. If the atom is in the excited state, $C_e = 1$ and $C_g = 0$, and the population inversion is 1, and the dipole moment is zero. If the atom is in the ground state, the population inversion is -1 and the dipole is again zero. The dipole is maximized when the atom is halfway between the excited and ground states, maximizing the product $C_e^* C_g$, on the equator of the Bloch sphere.

The equations for the expectation values of the Pauli operators could have been obtained from the $\dot{C}$ equations via

$$\langle\dot{\sigma}_z\rangle = \dot{C}_e C_e^* + \dot{C}_e^* C_e + \dot{C}_g C_g^* + \dot{C}_g^* C_g \tag{4.145}$$

$$\langle\dot{\sigma}_+\rangle = \dot{C}_e^* C_g + C_e^* \dot{C}_g \tag{4.146}$$

$$\langle\dot{\sigma}_-\rangle = \dot{C}_g^* C_e + C_g^* \dot{C}_e \tag{4.147}$$

We now introduce damping in a phenomenological manner, with a decay rate γ for the population inversion, and Γ for the dipole,

$$\frac{d\Sigma_+}{dt} = (i\Delta - \Gamma)\Sigma_+ - \frac{i\Omega_0}{2}\Sigma_z \tag{4.148}$$

$$\frac{d\Sigma_-}{dt} = -(i\Delta + \Gamma)\Sigma_+ - \frac{i\Omega_0}{2}\Sigma_z \tag{4.149}$$

$$\frac{d\Sigma_z}{dt} = -\gamma(\Sigma_z + 1) + i\Omega_0(\Sigma_+ - \Sigma_-) \tag{4.150}$$

The reason for the $\Sigma_z + 1$ is that in the absence of a driving field, we require that the population inversion decays to -1, that is the atom in the ground state. We also have $\Gamma = \gamma/2 + \gamma_{\text{phase}}$, where γ_{phase} is a rate of decay of the dipole that comes from a process that disturbs the phase of the dipole randomly, but does not affect the population inversion, for example, collisions between atoms in a vapor cell. The dipole decays at half the rate of the population inversion if only radiative decay (spontaneous emission) is involved. The intensity of spontaneously emitted light is proportional to the population inversion and decays as $\exp(-\gamma t)$. The radiated field is proportional to the dipole moment and should decay at half the rate, as the field is the square root of the intensity. These equations are known as the optical Bloch equations.

These equations have the following steady state solutions for expectation values:

$$\langle \Sigma_+ \rangle^{ss} = -\frac{\Omega}{2\Omega_{\text{sat}}}\sqrt{\frac{\gamma}{\Gamma}}\frac{(1 - i\tilde{\Delta})}{(1 + \tilde{\Delta}^2 + \Omega^2/\Omega_{\text{sat}}^2)} \tag{4.151}$$

$$\langle \Sigma_- \rangle^{ss} = -\frac{\Omega}{2\Omega_{\text{sat}}}\sqrt{\frac{\gamma}{\Gamma}}\frac{(1 + i\tilde{\Delta})}{(1 + \tilde{\Delta}^2 + \Omega^2/\Omega_{\text{sat}}^2)} \tag{4.152}$$

$$\langle \Sigma_z \rangle^{ss} = \frac{-(1 + \tilde{\Delta}^2)}{(1 + \tilde{\Delta}^2 + \Omega^2/\Omega_{\text{sat}}^2)} \tag{4.153}$$

where we have defined $\tilde{\Delta} = \Delta/\Gamma$ and $\Omega_{\text{sat}}^2 = \gamma\Gamma$. These are a dimensionless detuning and a saturation Rabi frequency. The latter is usually expressed as a saturation field, $E_{\text{sat}} = \sqrt{\gamma\Gamma}\,\hbar/\mu_{eg}$, or a saturation intensity $I_{\text{sat}} = \varepsilon_0 c E_{\text{sat}}^2$. Plots of the atomic dipole and population inversion as a function of driving field are shown in figures 4.6 and 4.7.

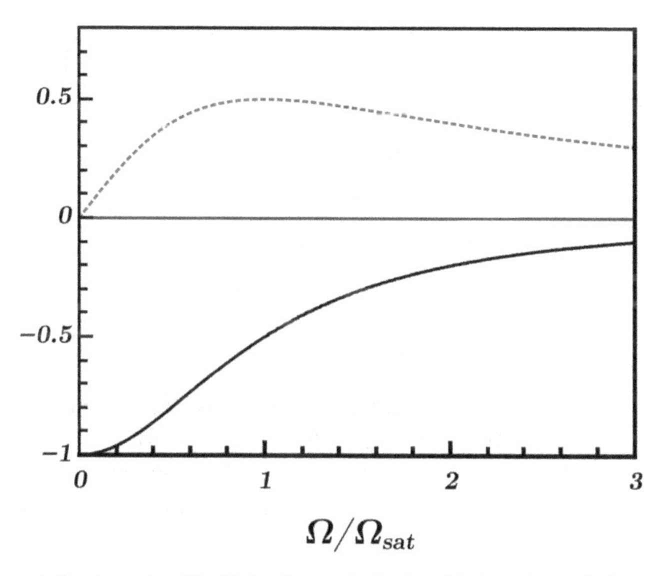

Figure 4.6. The population inversion (black), in-phase polarization (blue), and out-of-phase polarization (red), as functions of driving intensity, for zero detuning.

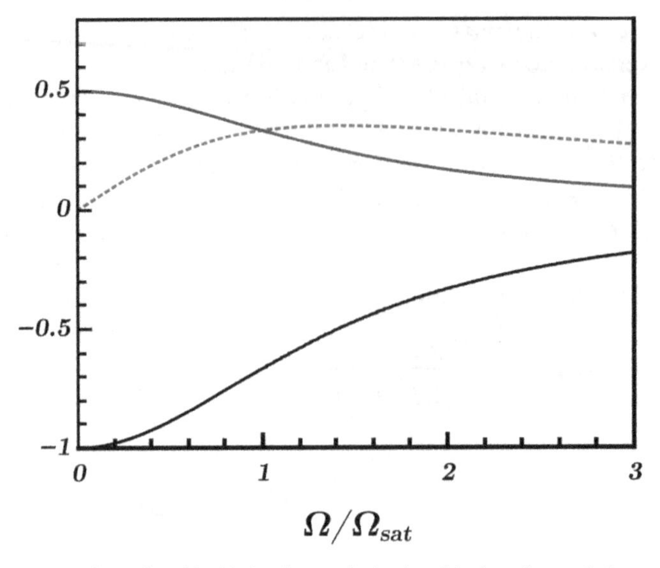

Figure 4.7. The population inversion (black), in-phase polarization (blue), and out-of-phase polarization (red), as functions of driving intensity, for nonzero detuning. Note the in-phase polarization is nonzero.

At the saturation intensity, the population inversion is halfway between the values at zero driving intensity and infinite driving intensity.

We see from these plots that the two-level atom is not a linear system, as the polarization is not linear with the driving field, but saturates. Generally in a linear, isotropic, homogeneous system, one has $P = \varepsilon_0 \chi E = N\mu_{eg}E$, where P is the polarization and χ is the permitivity of the material. In an anisotropic material, $\chi \to \vec{\chi}$, and the polarization does not lie in the same direction as the applied field, an inhomogeneous material would have a positional dependence for $\chi = \chi(\vec{r})$. For a nonlinear material, χ is not a constant, but is a function of the driving field. For our two-level atom, we have (on resonance)

$$\chi = \frac{N\mu_{eg}/\hbar\varepsilon_0\Omega_{\text{sat}}}{1 + \Omega^2/\Omega_{\text{sat}}^2} \tag{4.154}$$

As we can see, this is a nonlinear susceptibility which can be expanded as

$$\begin{aligned}\chi &= N\mu_{eg}/\hbar\varepsilon_0\Omega_{\text{sat}} - N\mu_{eg}^3/\hbar^3\varepsilon_0\Omega_{\text{sat}}^3 + \cdots \\ &= \chi^{(1)} + \chi^{(3)} + \cdots\end{aligned} \tag{4.155}$$

The nonlinearity is cubic to first order, as it should be for a medium with inversion symmetry, as a collection of atoms typically is (figure 4.8).

Another form of the Bloch equations involves the so-called in- and out-of-phase quadratures $\sigma_x = U\cos(\omega t) - V\sin(\omega t)$. These are related to the polarization of course via

$$P = N\mu_{eg}\sigma_x \tag{4.156}$$

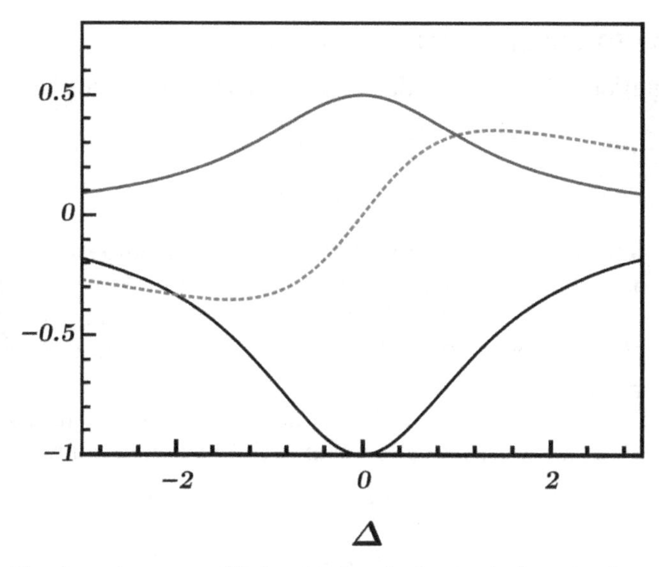

Figure 4.8. The absorption spectra (blue) and index of refraction (red), as functions of detuning.

These quadratures are defined via

$$U = \langle\Sigma_+\rangle + \langle\Sigma_-\rangle \tag{4.157}$$

$$V = -i(\langle\Sigma_+\rangle - \langle\Sigma_-\rangle) \tag{4.158}$$

$$W = \langle\Sigma_z\rangle \tag{4.159}$$

These satisfy the following dynamical equations:

$$\dot{U} = \Delta V - \Gamma U \tag{4.160}$$

$$\dot{V} = -\Delta U + \Omega W - \Gamma V \tag{4.161}$$

$$\dot{W} = -\gamma(W + 1) - \Omega V \tag{4.162}$$

The physical interpretation of these is the population inversion (W), the dipole moment $\pi/2$ out of phase with the driving field, scaled by μ_{eg} (V), and the dipole moment in phase with the driving field (U). On resonance, the steady state dipole is $\pi/2$ out of phase with the driving field, as a damped driven simple harmonic oscillator is. In terms of latitude and longitude on the Bloch sphere, we have

$$W = \cos\theta \tag{4.163}$$

$$U = \sin\theta\sin\phi \tag{4.164}$$

$$V = \sin\theta\cos\phi \tag{4.165}$$

and without decay processes, we have $U^2 + V^2 + W^2 = 1$, the equation for the Bloch sphere itself.

4.6 Relation to the classical Lorentz model

For weak excitation, we can approximate $W = -1$ and find equations for U and V as

$$\dot{U} = \Delta V - \Gamma U \tag{4.166}$$

$$\dot{V} = -\Delta U - \Omega - \Gamma V \tag{4.167}$$

These are the same equations that result from the Lorentz model, where we assume that the electron is bound to the nucleus by a spring of $k = m\omega_0^2$, as we saw in chapter 2, driven by an external field $E_0 \cos \omega t$, with damping rate Γ. The Rabi frequency $\mu_{eg}E/\hbar$ is analogous to what we referred to as a classical Rabi frequency in chapter 1, $\Omega_{cl} = eE/2m\omega X_0$.

We can combine the equations for U, V, and W with field equations derived from Maxwell's equations to write down the Maxwell–Bloch equations,

$$\dot{U} = \Delta V - \Gamma U \tag{4.168}$$

$$\dot{V} = -\Delta U + \Omega W - \Gamma V \tag{4.169}$$

$$\dot{W} = -\gamma(W + 1) - \Omega V \tag{4.170}$$

$$\dot{A} = -\kappa A - \frac{1}{2}\omega V \tag{4.171}$$

$$\dot{\phi} = \Omega - \omega - \frac{1}{2}\omega U \tag{4.172}$$

where A is the amplitude of the field and ϕ is the phase. If the field is a constant applied field, then of course we only need the Bloch equations. The Maxwell–Bloch equations are used to describe an atom or atoms inside an optical cavity. Here, A would be the amplitude of the intracavity field. Usually, there is a driving field representing a laser input to one of the end mirrors, the Maxwell–Bloch equations would then be modified to read

$$\dot{U} = \Delta V - \Gamma U \tag{4.173}$$

$$\dot{V} = -\Delta U + \Omega W - \Gamma V \tag{4.174}$$

$$\dot{W} = -\gamma(W + 1) - \Omega V \tag{4.175}$$

$$\dot{A} = -\kappa A - \frac{1}{2}\omega V + Y \tag{4.176}$$

$$\dot{\phi} = \Omega - \omega - \frac{1}{2}\omega U \tag{4.177}$$

where Y is the amplitude of the driving field multiplied by the transmission of the mirror the driving field is incident on, and we assume no phase shift for transmission through that mirror.

4.7 An interlude in the form of the atom–field interaction

The standard way to introduce atom–field interactions into quantum field theory is the minimal coupling replacement

$$\vec{p} \rightarrow \vec{\pi} = \vec{p} - \frac{e}{c}\vec{A} \tag{4.178}$$

where $\vec{A}$ is the usual vector potential. The quantity $\vec{\pi}$ is the canonical momentum that has the operator representation $\vec{\pi} = -i\hbar\vec{\nabla}$. The usual kinetic energy term in the Hamiltonian then becomes

$$\begin{aligned}\frac{\pi^2}{2m} &\rightarrow \frac{1}{2m}\left(\pi^2 + \frac{e^2}{c^2}|A|^2 - \frac{e}{c}\vec{A}\cdot\vec{\pi} - \frac{e}{c}\vec{\pi}\cdot\vec{A}\right)\\ &= \frac{1}{2m}\left(\pi^2 + \frac{e^2}{c^2}|A|^2 - 2\frac{e}{c}\vec{A}\cdot\vec{\pi}\right)\end{aligned} \tag{4.179}$$

where we have used the Coulomb gauge condition $\vec{\nabla}\cdot\vec{A} = 0$ to show that in that gauge the vector potential and the momentum commute

$$\begin{aligned}\vec{\pi}\cdot\vec{A}|\psi\rangle &= -i\hbar\vec{\nabla}\cdot\vec{\pi}|\psi\rangle\\ &= -i\hbar(\vec{\nabla}\cdot\vec{A})|\psi\rangle - i\hbar\vec{A}\cdot\vec{\nabla}|\psi\rangle\\ &= -i\hbar\vec{A}\cdot\vec{\nabla}|\psi\rangle\\ &= \vec{A}\cdot\vec{\pi}|\psi\rangle\end{aligned} \tag{4.180}$$

which leads us to an interaction Hamiltonian of the form

$$H_{\text{int}} = -\frac{e}{m}\vec{A}\cdot\vec{\pi} + e^2|A|^2 \tag{4.181}$$

The spatial dependence of components of $\vec{A}$ for a plane electromagnetic wave of wavevector $\vec{k}$ is

$$A_i = A_i(t)e^{i\vec{k}\cdot\vec{r}} \tag{4.182}$$

In the dipole approximation that the field does not vary much over a wavelength of light, we can evaluate the exponential factor at the location of the atom which we take to be $\vec{r} = 0$, then the components of $\vec{A}$ have a time dependence but no spatial dependence.

We now consider a gauge transformation [1, 2]

$$\vec{A}' = \vec{A} - \vec{\nabla}\lambda \tag{4.183}$$

$$\phi' = \phi + \frac{\partial\lambda}{\partial t} \tag{4.184}$$

with the choice

$$\lambda = \vec{A}\cdot\vec{r} \tag{4.185}$$

Of course, in quantum mechanics, when one performs a gauge transformation on the vector and scalar potentials, there is an accompanying transformation of the wavefunction,

$$|\psi'\rangle = e^{ie\lambda/\hbar c}|\psi\rangle \tag{4.186}$$

The Schrödinger equation for the new wave function $|\psi'\rangle$ is

$$\begin{aligned} i\hbar|\dot{\psi}\rangle &= i\hbar\left(\frac{ie\dot{\lambda}}{\hbar c}|\psi'\rangle + |\dot{\psi}'\rangle\right)e^{ie\vec{A}\cdot\vec{r}/\hbar c} \\ &= i\hbar\left(\frac{ie\dot{\vec{A}}\cdot\vec{r}}{\hbar c}|\psi'\rangle + \frac{ie\vec{A}\cdot\dot{\vec{r}}}{\hbar c}|\psi'\rangle + |\dot{\psi}'\rangle\right)e^{ie\vec{A}\cdot\vec{r}/\hbar c} \\ &= e^{ie\vec{A}\cdot\vec{r}/\hbar c}\left(\frac{\pi^2}{2m} + V(r)\right)|\psi'\rangle \end{aligned} \tag{4.187}$$

Using $\vec{E} = -\dot{\vec{A}}$, and the Lorentz gauge $\vec{\nabla}\cdot\vec{A} = 0$, we find

$$i\hbar|\dot{\psi}'\rangle = \left(\frac{\pi^2}{2m} + V(r) - e\vec{r}\cdot\vec{E}\right)|\psi'\rangle \tag{4.188}$$

So, we can obtain the electric dipole form from the canonical form. This has generated a periodic debate in the literature, with claims that the two forms give different answers, the typical resolution is that someone has forgotten to use $|\psi'\rangle$ with the $H_I = -\vec{\mu}\cdot\vec{E}$, instead using $|\psi\rangle$ or the other way around. The discrepancy between the two 'different' answers oftentimes is of the form of some factor of ω/ω_0 which is oftentimes small for optical problems, nevertheless care should be taken! The advantage of the electric dipole form is that the field is gauge invariant and this Hamiltonian includes the A^2 term. Conversely, going beyond the dipole approximation is easy in the $\vec{A}\cdot\vec{\pi}$ form by using the spatial dependence of the vector potential, in the other approach one must use a multipolar expansion. Both forms yield equivalent results if one is careful, but when mixed with other approximations like the rotating wave approximation and treating 'real' systems as two-level atoms, differences may arise.

References

[1] Milonni P 2019 *An Introduction to Quantum Optics and Quantum Fluctuations* (Oxford: Oxford University Press)

[2] Milonni P W 1994 *The Quantum Vacuum: An Introduction to Quantum Electrodynamics* (New York: Academic)

IOP Publishing

An Introduction to Quantum Optics

An open systems approach

Perry Rice

Chapter 5

Quantum fields

5.1 Maxwell equations again

Recall that these are

$$\vec{\nabla} \cdot \vec{E} = \frac{\rho}{\varepsilon_0} \tag{5.1}$$

$$\vec{\nabla} \cdot \vec{B} = 0 \tag{5.2}$$

$$\vec{\nabla} \times \vec{E} = \frac{-\partial \vec{B}}{\partial t} \tag{5.3}$$

$$\vec{\nabla} \times \vec{B} = \mu_0 \vec{J} + \frac{1}{c^2}\frac{\partial \vec{E}}{\partial t} \tag{5.4}$$

Also, we have introduced the vector potential $\vec{A}$ via $\vec{B} = \vec{\nabla} \times \vec{A}$, and the scalar potential ϕ via $\vec{E} = -\vec{\nabla}\phi - \partial\vec{A}/\partial t$.

5.2 Quantization of the electromagnetic field

Let us now explicitly prove show how the EM field becomes quantized, and the resultant effects on the cavity field modes. We follow the method used by Loudon [5] and Cohen-Tannoudji *et al* [2, 3]. We start with the equation for the vector potential developed in chapter 3

$$-\nabla^2\vec{A} + \frac{1}{c^2}\frac{\partial^2\vec{A}}{\partial t} = \mu_0\vec{J}_T \tag{5.5}$$

We consider free space where $\vec{J}_T = \rho = \phi = 0$. The wave equation for $\vec{A}$ is then

$$-\nabla^2 \vec{A} + \frac{1}{c^2}\frac{\partial^2 \vec{A}}{\partial t^2} = 0 \tag{5.6}$$

The solution to the wave equation for $\vec{A}$ using a plane wave expansion is

$$\vec{A} = \sum_{k,l}\left(A_{k,l}(0)e^{i(\vec{k}\cdot\vec{r}-\omega_k t)} + A^*_{k,l}(0)e^{-i(\vec{k}\cdot\vec{r}-\omega_k t)}\right)\hat{\varepsilon}_{k,l} \tag{5.7}$$

where $k_x = 2\pi n_x/L$ and similarly for k_y and k_z. These are appropriate choices for box quantization, where we have assumed that the field must vanish at the boundaries of a cube of sides L. For a rectangular cavity, one would simply use different lengths for each k_i, and of course other geometries are possible. Later, we will give some details for real cavities. The sum over k is a sum over all directions in space, all possible frequencies $\omega_k = ck$, and also over all polarizations. Now, since $\vec{\nabla} \cdot \vec{A} = 0$, this means that $\vec{k} \cdot \vec{A}_k = \vec{k} \cdot \hat{\varepsilon}_{k,l} = 0$. The two orientations for the polarization are two unit vectors $\hat{\varepsilon}_{k1}$ and $\hat{\varepsilon}_{k2}$, which along with $\vec{k}$, form a right-handed set of orthogonal vectors for each $\vec{k}$. As $\vec{A}$ is driven by the transverse part of the current density $\vec{J}_T$, we only have transverse solutions. Instead of plane waves, we could use an expansion in terms of cylindrical or spherical waves; we could also use complex polarization vectors $\hat{\varepsilon}_{R,L} = (1/\sqrt{2})(\hat{\varepsilon}_{k,1} \pm \hat{\varepsilon}_{k,2})$ for right and left circular polarizations.

The electric field is then given by

$$\begin{aligned}\vec{E} &= -\frac{\partial \vec{A}}{\partial t}\\ &= \sum_{k,l} - i\omega_k\left(A_{k,l}(0)e^{i(\vec{k}\cdot\vec{r}-\omega_k t)} - A^*_{k,l}(0)e^{-i(\vec{k}\cdot\vec{r}-\omega t)}\right)\\ &= \sum_{k,i} E_{k,l}\hat{\varepsilon}_{k,i}\end{aligned} \tag{5.8}$$

with

$$E_{k,l} = -i\omega_k A_{k,l}(0)e^{i(\vec{k}\cdot\vec{r}-\omega_k t)} + c.c., \tag{5.9}$$

where $c.c.$ stands for complex conjugate. The magnetic field is then given by

$$\vec{B} = \sum_{k,i} B_{k,l}(\hat{k} \times \hat{\varepsilon}_{k,l}) \tag{5.10}$$

with

$$B_{k,l} = ikA_{k,l}e^{i(\vec{k}\cdot\vec{r}-\omega_k t)} + c.c. \tag{5.11}$$

We see that $|B_k| = \frac{1}{c}|E_k|$, so then the energy of the kth mode time averaged (denoted by the bar) is

$$\overline{\varepsilon_k} = \frac{1}{2}\int\left(\varepsilon_0 \bar{E}_k^2 + \frac{1}{\mu_0}\bar{B}_k^2\right)dV$$
$$= 2\varepsilon_0 V \omega_k^2 \vec{A}_k \cdot \vec{A}_k^* \tag{5.12}$$

Now, we transform the above equations into a harmonic oscillator with polarization vectors $\vec{\varepsilon}_k$ so that the vector potential and its complex conjugate can be written as

$$\vec{A}_{k,l} = (4\pi\varepsilon_0 V \omega_k^2)^{1/2}(\omega_k Q_{k,l} + iP_{k,l})\hat{\varepsilon}_{k,l} \tag{5.13}$$

$$\vec{A}_{k,l}^* = (4\pi\varepsilon_0 V \omega_k^2)^{1/2}(\omega_k Q_{k,l} - iP_{k,l})\hat{\varepsilon}_{k,l} \tag{5.14}$$

The time average of the energy of a given mode is then

$$\overline{\varepsilon_{k,l}} = \frac{1}{2}\left(P_{k,l}^2 + \omega_k^{2,}l2\right) \tag{5.15}$$

For a quantum harmonic oscillator, we know that the Hamiltonian is

$$H = \frac{1}{2}(p^2 + \omega^2 q^2) \tag{5.16}$$

and we know that the appropriate commutation relation is

$$[q, p] = -i\hbar \tag{5.17}$$

So, as usual, let us define a set of operators that obey the following:

$$\hat{a} = (2\hbar\omega)^{1/2}(\omega q + ip) \tag{5.18}$$

$$\hat{a}^\dagger = (2\hbar\omega)^{-1/2}(\omega q - ip) \tag{5.19}$$

$$q = \left(\frac{\hbar}{2\omega}\right)^{1/2}(a + a^\dagger) \tag{5.20}$$

$$p = i\left(\frac{\hbar\omega}{2}\right)^{1/2}(a - a^\dagger) \tag{5.21}$$

which satisfy the commutation relation

$$[a, a^\dagger] = 1 \tag{5.22}$$

Consider the product $a^\dagger a$, we find that

$$a^\dagger a = \frac{1}{2\hbar\omega}(p^2 + \omega^2 q^2 + i\omega qp - i\omega pq)$$
$$= \frac{1}{\hbar\omega}\left(H - \frac{1}{2}\hbar\omega\right) \tag{5.23}$$

which leaves us with

$$H = \hbar\omega\left(a^\dagger a + \frac{1}{2}\right) \tag{5.24}$$

Let us consider energy eigenstates, or number states $|n\rangle$. Here, number refers to the number of photons in the mode

$$\begin{aligned} H|n\rangle &= E_n|n\rangle \\ &= \left(a^\dagger a + \frac{1}{2}\right)\hbar\omega|n\rangle \end{aligned} \tag{5.25}$$

What is the effect of the operator $a^\dagger$ on eigenstates $|n\rangle$. We act on $a^\dagger|n\rangle$ with H

$$\begin{aligned} \hbar\omega\left(a^\dagger a a^\dagger + \frac{1}{2}a^\dagger\right)|n\rangle &= \hbar\omega\left(a^\dagger a^\dagger a + a^\dagger + \frac{1}{2}a^\dagger\right)|n\rangle \\ &= \hbar\omega a^\dagger\left(a^\dagger a + \frac{3}{2}\right)r|n\rangle \\ &= a^\dagger(H + \hbar\omega)|n\rangle \\ &= (E_n + \hbar\omega)a^\dagger|n\rangle \end{aligned} \tag{5.26}$$

We see that $a^\dagger|n\rangle$ is also an energy eigenstate with $E = E_n + \hbar\omega$. This means that $a^\dagger$ raises the state to an energy level $\hbar\omega$ higher, and we refer to $a^\dagger$ as a raising operator. Likewise, it can be easily shown a is a lowering operator represented by

$$H(a|n\rangle) = (E_n - \hbar\omega)(a|n\rangle) \tag{5.27}$$

As the lowest state is defined to be $|0\rangle$, we know that $a|0\rangle = 0$, so that we can find out the ground state energy

$$\begin{aligned} H|0\rangle &= \hbar\omega\left(a^\dagger a + \frac{1}{2}\right)|0\rangle \\ &= \frac{1}{2}\hbar\omega|0\rangle \\ &= E_0|0\rangle \end{aligned} \tag{5.28}$$

This yields the ground state energy $E_0 = \hbar\omega/2$. Another state, $|1\rangle$, with energy $E_1 = 3\hbar\omega/2$ is obtained by raising the ground state. Are there any states between $|0\rangle$ and $|1\rangle$? Our notation would suggest not, and indeed there is none. If there was a state with energy $\hbar\omega/2 < E < 3\hbar\omega/2$, we could lower it and obtain a state with energy below that of the ground state. Since by construction the ground state is the state of lowest energy, we would have a contradiction. So, we know that our energy eigenstates are equally spaced, with energies $E_n = \hbar\omega(n + 1/2)$. We note that the ground state energy is not zero, the energy $E_0 = \hbar\omega/2$ is referred to as the zero-point energy.

Let us now look at exactly how the raising and lowering operators act on our number states.

$$\begin{aligned}a|n\rangle &= C_n|n-1\rangle \\ \langle n|a^\dagger &= \langle n-1|C_n^* \\ \langle n|a^\dagger a|n\rangle &= |C_n|^2\langle n-1|n-1\rangle \\ &= |C_n|^2 \\ &= n \\ C_n &= \sqrt{n}\end{aligned} \tag{5.29}$$

Similarly, we find that

$$a^\dagger|n\rangle = \sqrt{n+1}\,|n+1\rangle \tag{5.30}$$

So, how do we quantize the field? We simply let the momentum P_k that was originally a scalar go to $\hat{p}_k$ which is an operator, as well as replacing Q_k by $\hat{q}_k$. Then, $A_k \rightarrow \hat{a}_{k,l}$ and $A_{k,l}^* \rightarrow \hat{a}_{k,l}^\dagger$.

This will give us the quantized electric and magnetic fields that are still given by the sums equations (5.8) and (5.10). with mode coefficients that are operators,

$$E_{k,l} = -i\left(\frac{\hbar\omega_k}{2\varepsilon_0 V}\right)^{1/2}\left[a_{k,l}e^{i(\vec{k}\cdot\vec{r}-\omega t)} - a_{k,l}^\dagger e^{-i(\vec{k}\cdot\vec{r}-\omega t)}\right] \tag{5.31}$$

$$B_k = \frac{i}{c}\left(\frac{\hbar\omega_k}{2\varepsilon_0 V}\right)^{1/2}\left[a_k e^{i(\vec{k}\cdot\vec{r}-\omega t)} - a_k^\dagger e^{-i(\vec{k}\cdot\vec{r}-\omega t)}\right] \tag{5.32}$$

5.3 Single mode quantized fields

Let us focus on just one mode of the electric field for now. How can I create a situation in the lab where the use of a single mode is appropriate? In general, one needs to consider waves propagating in all directions in space, at all wavelengths/frequencies, and two polarizations. The simplest manner to reduce the number of modes is to use a pair of mirrors to make an optical cavity. For mirrors made from a perfect conductor, the boundary conditions are such that the field must vanish at the mirrors. As we stated before, this gives rise to a quantization of wavenumbers for modes in the cavity, $k_n = 2\pi nc/L$, that denote what we refer to as *longitudinal modes*. These have a spatial dependence of $U_n = \cos(kz)$ or $\sin(kz)$, where we have taken the z-axis perpendicular to the cavity mirrors. The transverse $x - y$ dependence is determined by a solution to Maxwell's equations, and various transverse modes emerge, typically the TEM (transverse–electromagnetic) modes labeled TEM_{ij}. The structure of these modes, for a rectangular cavity are proportional to Hermite polynomials, and some representative patterns are shown below. The losses for the higher order modes are typically larger than those of the TEM_{00}, or that can be arranged with an intracavity element like a pinhole. Also, a polarizing element may be placed in the cavity to allow only one polarization to propagate. The mirrors are typically curved, if for no other reason than to facilitate alignment and to keep the beam from walking out of the cavity. The $1/e$ points of the transverse beam profile define what is referred to as the waist of the beam. If we place atoms in our cavity that have a linewidth small compared to the spacing between longitudinal modes, we can restrict ourselves to consideration of a single mode. Now of course there are modes that propagate parallel to the mirrors, which an atom might spontaneously emit into. This can be suppressed if we have a cavity structure in those directions that is shorter than $\lambda/2$. There are various other geometries one can consider like a microdisk. Here, the thickness is less than $\lambda/2$ which will suppress coupling to modes perpendicular to the disk, and total internal reflection confines modes that propagate around the disk, so-called whispering gallery modes. The grazing incidence of these waves and the index mismatch between the semiconductor and air can lead to relatively large reflectivities, and hence cavity quality factors. But for any cavity, some light is transmitted. Thus our field couples to other modes, and we will consider that in more detail when we consider open quantum systems in chapter 8.

For our mathematical representation of a single mode field, we take

$$E_{k,l} \rightarrow E = \mathcal{E}_0(ae^{i\theta} + a^{\dagger}e^{-i\theta}) \tag{5.33}$$

where we have absorbed a factor of $i = e^{i\pi/2}$ into θ for convenience, and $\mathcal{E}_0 = \sqrt{\hbar\omega/2\varepsilon_0 V}$, where V is the volume of the cavity mode. Absorbing the factor of i is not a problem as the absolute phase is not measurable, only relative phases. The cavity mode volume is formally defined by setting the energy stored in the field equal to $\varepsilon_0|E_{\max}|^2 V_{\text{eff}}$, or

$$V_{\text{eff}} = \frac{1}{|E_{\max}|^2} \int |\vec{E}(\vec{r})|^2 dV \tag{5.34}$$

For a sinusoidal mode with a Gaussian profile,

$$E = E_{\max}\sin(kz)e^{-(x^2+y^2)/w_0^2} \tag{5.35}$$

where w_0 is the beam waist, and the effective cavity mode volume is

$$V = \frac{\pi w_0^2}{4}L \tag{5.36}$$

where we have assumed that the beam waist is constant (oftentimes it is not due to the hourglass shape of some modes) and that the field vanishes at the mirrors spaced a length L apart. In a semiconductor material, or any material with an index, one must take the index of refraction into account, $V \rightarrow V/n^3$, or one must include the spatial dependence of the dielectric constant in a more complicated cavity, perhaps composed of a photonic band gap material. More problematic may be that most mirrors are not perfect, or even imperfect conductors where the field nearly vanishes at the mirror. Most real mirrors, that are used to get large reflectivities, use interference effects to obtain large reflectivities at a particular narrow range of frequencies. These mirrors are alternating layers of a dielectric, and the field mode can penetrate into that dielectric stack. So, one must be more careful not to use the above formula incautiously in that type of situation. Generally, one must model the mirror, and solve for the true classical modes, and then integrate that up to determine the effective volume. If for a cavity of dielectric mirrors one uses that formula for conducting mirrors, and uses the spacing between the top layers of the stack, one will be using a volume that is too small, which results in overestimating that electric field per photon.

Oftentimes it is useful to consider the field written as a sum of quadratures,

$$E = \mathcal{E}_0(X\cos(\theta) + Y\sin(\theta)) \tag{5.37}$$

where the quadrature operators are defined by

$$X = \frac{1}{2}(a^\dagger + a) \tag{5.38}$$

$$Y = \frac{i}{2}(a^\dagger - a) \tag{5.39}$$

The energy per volume in this mode is given by

$$\bar{\varepsilon} = \frac{\varepsilon\mathcal{E}_0^2}{2}(X^2 + Y^2) \tag{5.40}$$

and so it is appropriate to refer to $\mathcal{E}_0$ as the electric field per photon. In some small cavities of a micron in length, $\mathcal{E}_0$ can be 1000–10,000 V/m, which is similar to the electric field across an AC power outlet, 100 volts dropped over roughly a centimeter. Thus the presence or absence of a single photon can change the electric field in a cavity by a significant amount, that is definitely macroscopic and measurable. We will say more about this in chapter 9 when we discuss quantum trajectory theory.

Recall that the Hamiltonian for an electromagnetic field mode resembles that of a harmonic oscillator,

$$\frac{p^2}{2m} + \frac{kx^2}{2} \leftrightarrow \frac{\varepsilon E^2}{2} + \frac{B^2}{2\mu_0} \tag{5.41}$$

For the harmonic oscillator, we have an uncertainty principle

$$\begin{aligned} \Delta x \Delta p &\geqslant \frac{1}{2}|\langle[x, p]\rangle| \\ &\geqslant \frac{\hbar}{2} \end{aligned} \tag{5.42}$$

It would seem that there might be an uncertainty principle between the electric and magnetic fields. How does this square with the fact that for a plane wave mode, $|B| = |E|/c$? We must recall that when we quantize classical objects and promote a classical c-number to an operator (say q), we can identify the variable conjugate to q which we denote by p. Then, we would have the relation $[q, p] = i\hbar$. In the case of electromagnetic fields, E and B are not conjugate variables, but E and A are, recall that $E \propto \dot{A}$ as $p \propto \dot{x}$. These are both functions of x, and as such their commutation relation is of the form

$$[E(x), A(x')] \propto \delta_T(x - x') \tag{5.43}$$

where $\delta_T(x - x')$ is a transverse delta function, related to the transverse nature of the vector potential. Let us continue to work with the quadrature operators X and Y. These operators have a commutation relation given by

$$[X, Y] = \frac{-i}{2} \tag{5.44}$$

which leads us to the uncertainty relation

$$\Delta X \Delta Y \geqslant \frac{1}{4} \tag{5.45}$$

For a classical field, X and Y could take on a well defined value, effectively a point in a type of phase space. For example, a field given by $E = X\cos(\omega t) + Y\sin(\omega t)$ would be described as a phasor. For a quantum field, we must replace that phasor, by a phasor for the mean field plus a type of quantum 'fuzzball'. For now the 'fuzzball' is a schematic representation of the uncertainty, or (we will use the term noise) in the state. This 'fuzzball' will take on formal meaning when we discuss quasi-probability amplitudes in chapter 10. There we will see that we can do quantum optics in a manner like classical statistical mechanics in phase space, using probability distributions. However, these distributions will have some bizarre properties, such as being negative. The quantum weirdness must show up somehow!

5.4 Number states

In a number state, or Fock state, we can show that

$$\langle X \rangle = 0 \tag{5.46}$$

$$\langle X^2 \rangle = (1/2)(n + 1/2) \tag{5.47}$$

$$\langle Y \rangle = 0 \tag{5.48}$$

$$\langle Y^2 \rangle = (1/2)(n + 1/2) \tag{5.49}$$

which leads to

$$\Delta X \Delta Y = (1/2)(n + 1/2) \geqslant 1/4 \tag{5.50}$$

where the minimum uncertainty occurs for the vacuum state $|0\rangle$. The ground state of a harmonic oscillator is a Gaussian in the position representation, which is a minimum uncertainty state. Pictorially, we may illustrate the number state in figure 5.1, where it seems that the amplitude of the field is perfectly well known, and the phase is completely unknown. Also, as we have

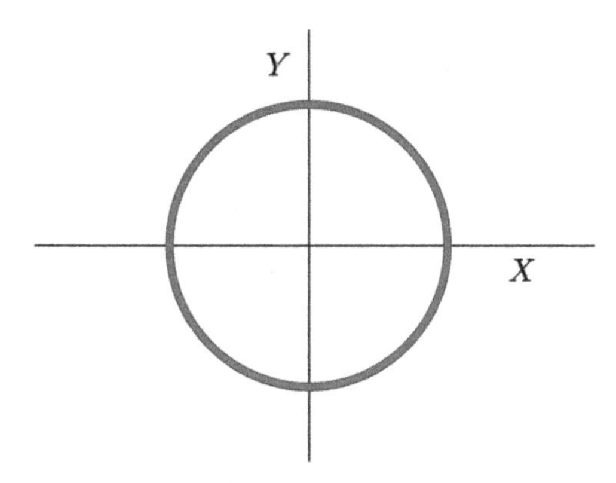

Figure 5.1. Cartoon of the quadrature distribution of a Fock state.

$$X^2 + Y^2 = a^\dagger a + 1/2 = N + 1/2 \tag{5.51}$$

where we use the notation $N \equiv a^\dagger a$ to distinguish it from its eigenvalues n, the number of photons in the Fock state. The Fock states obviously satisfy

$$N|n\rangle = n|n\rangle \tag{5.52}$$

$$N^2|n\rangle = n^2|n\rangle \tag{5.53}$$

which then leads to $\Delta N = 0$ and of course the photon number is perfectly known.

This leads us to consider the question of amplitude and phase operators. We can write the electric field as

$$E = A\, exp(i\phi) \tag{5.54}$$

with A and ϕ Hermitian operators that can then be measured? The answer is no. Dirac initially considered operators of this form where $A = \sqrt{N}$, and the commutation relations for these operators is

$$[e^{i\phi}, N] = e^{i\phi} \tag{5.55}$$

This would be satisfied if we have

$$[N, \phi] = \frac{i}{2} \tag{5.56}$$

which would yield an uncertainty relation

$$\Delta N \Delta\phi \geqslant \frac{1}{2} \tag{5.57}$$

So far, this seems OK, but what if we consider

$$\langle n|[N, \phi]|n'\rangle = i\delta_{n,n'} \tag{5.58}$$

but if we let the number operator N act on the states, we find

$$(n - n')\langle n|\phi|n'\rangle = i\delta_{n,n'} \tag{5.59}$$

which leads to a contradiction for $n = n'$, unless the matrix elements of the phase operator in the number basis do not exist, leading to the conclusion that ϕ cannot be a well-defined Hermitian operator. One can also show that the following relations hold:

$$(e^{i\phi})(e^{i\phi})^{\dagger} = 1 \tag{5.60}$$

$$(e^{i\phi})^{\dagger}(e^{i\phi}) \neq 1 \tag{5.61}$$

If ϕ were a Hermitian operator, then $e^{i\phi}$ should be unitary, and it is not.

The failure to define a Hermitian phase operator stems from the difficulty of defining a phase for the ground state $|0\rangle$, which (at our level of rigor so far) described by a circular 'fuzzball' around the origin [4]. The notion of a 'fuzzball' will be formalized in chapter 10 where we discuss quasiprobability distributions. Also, as for an oscillating field, we have $\phi = \omega t$ which would mean that if we *could* define a phase operator, we would also have a time operator. In nonrelativistic quantum mechanics of course, time is merely a parameter. More sophisticated ways to deal with time in quantum mechanics must and do show up in a relativistic formulation. There are various versions of amplitude and phase operators that work on some subset of states, even all states but the vacuum state, or operators that can be used to construct a probability distribution for phase that has the correct properties, but there is no rigorous amplitude/phase uncertainty relation. Susskind and Glogower [10] considered operators of the form

$$e = \widehat{e^{i\phi}} = (N + 1)^{-1/2}a \tag{5.62}$$

$$e^{\dagger} = \widehat{e^{-i\phi}} = a^{\dagger}(N + 1)^{-1/2} \tag{5.63}$$

These new operators are not unitary either as

$$e^{\dagger}e = 1 - |0\rangle\langle 0| \tag{5.64}$$

$$ee^{\dagger} = 1 \tag{5.65}$$

but there are useful operators

$$C = \frac{1}{2}(e + e^{\dagger}) \tag{5.66}$$

$$S = \frac{1}{2i}(e - e^{\dagger}) \tag{5.67}$$

with commutators

$$[C, N] = iS \tag{5.68}$$

$$[S, N] = -iC \tag{5.69}$$

which leads to uncertainties

$$\Delta N \Delta C \geqslant \frac{1}{2}|\langle S \rangle| \tag{5.70}$$

$$\Delta N \Delta S \geqslant \frac{1}{2}|\langle C \rangle| \tag{5.71}$$

More recent work by Pegg and Barnett [8], Mandle [7], and others [1] has helped formulate the problem of phase operator and phase distributions, with a focus on measurable quantities. A cartoon of the quadrature distribution of a Fock state is shown in figure 5.1.

The photon number distribution for a Fock state is rather boring, as we have a unitary probability for $n = N_0$, and zero otherwise, as shown in figure 5.2.

The time evolution of the photon number is shown in figure 5.3. The blue curve denotes the expectation value, while the red shaded region denotes the variance. There is no time dependence, other than the factor $e^{-in\omega_0 t}$ which has a magnitude of unity.

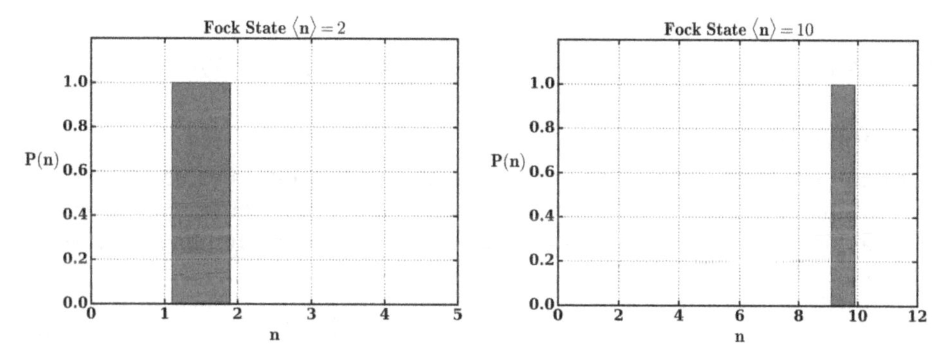

Figure 5.2. Photon number probability distribution for Fock states $|n = 2\rangle$ and $|n = 10\rangle$.

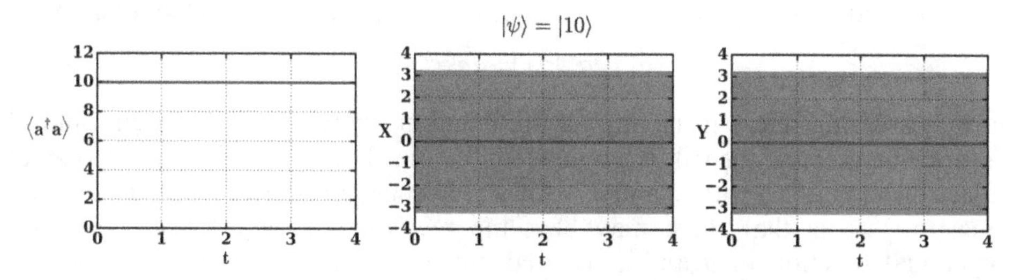

Figure 5.3. Time evolution of the average (blue curve) and variance (red shaded region) of a number state $|n = 10\rangle$.

Finally, we consider how to generate Fock states from the ground state. Using the properties of the raising operator, we have

$$\begin{aligned}|1\rangle &= a^\dagger|0\rangle\\ |2\rangle &= \frac{a^\dagger}{\sqrt{2}}|1\rangle\\ |3\rangle &= \frac{a^\dagger}{\sqrt{3}}|2\rangle\end{aligned} \tag{5.72}$$

We can see then how we can form

$$|n\rangle = \frac{a^{\dagger n}}{\sqrt{n!}}|0\rangle \tag{5.73}$$

Since any state can be decomposed into a superposition of Fock states, we can find an operator that generates any state from the ground state,

$$\begin{aligned}|\psi\rangle &= \sum_n C_n|n\rangle\\ &= \sum_n \frac{C_n a^{\dagger n}}{\sqrt{n!}}|0\rangle\\ &= S(C_n)|0\rangle\end{aligned} \tag{5.74}$$

where we define the state generating operator $S(C_n) = \sum_n \frac{C_n a^{\dagger n}}{\sqrt{n!}}|0\rangle$.

5.5 Coherent states

We begin by finding states that have a well-defined expectation value of the electric field that might possibly be related to a classical field. Consider $\langle E\rangle = \langle(a^\dagger + a)\rangle$. If we could find a state of the field that was an eigenstate of the annihilation operator, $a|\alpha\rangle = \alpha|\alpha\rangle$, then we would have

$$\langle 2X\rangle = \langle(a^\dagger + a)\rangle = (\alpha + \alpha^*) = 2\Re(\alpha) \tag{5.75}$$

$$\langle 2Y\rangle = i\langle(a^\dagger - a)\rangle = i(\alpha^* - \alpha) = 2\Im(\alpha) \tag{5.76}$$

where we have used the conjugate relation $\langle\alpha|a^\dagger = \langle\alpha|\alpha^*$. This state is even more special, consider the expectation value of any normally ordered product of a and $a^\dagger$,

$$\langle\alpha|a^{\dagger n}a^m|\alpha\rangle = \alpha^{*n}\alpha^m \tag{5.77}$$

Normal ordering refers to having all annihilation operators on the right, and all creation operators on the left. It would seem that for these types of products, one can replace the operator a with α, and $a^\dagger$ with α^*, replacing an operator with a c number. Thus, this seems that there may be some relation with a classical field. Any polynomial function of a and $a^\dagger$ can be put in the form of a sum of normally ordered products using the commutation relation $[a, a^\dagger] = 1$; and so we can always use these states, called coherent states, to calculate all quantities of interest. For example, the quantity $aa^\dagger + a^\dagger a = 2aa^\dagger a + 1$, and so for a coherent state we would have

$$\langle aa^\dagger + a^\dagger a\rangle = \langle 2a^\dagger a + 1\rangle = 2|\alpha|^2 + 1 \tag{5.78}$$

Of course for a classical field, we would not need to have the various extra factors due to the non-commutative nature of a and $a^\dagger$. So, while one can always use the coherent states for calculation, it is not always the case that operators are replaced by c-numbers, rather one must use the commutation relations first. Therefore, the quantum nature of the field can still 'peek' through.

How do we find these coherent states? Consider the expansion

$$|\alpha\rangle = \sum_n C_n|n\rangle \tag{5.79}$$

Using the definition of the coherent state, we have

$$\begin{aligned} a|\alpha\rangle &= \sum_n C_n a|n\rangle \\ &= \sum_n C_n\sqrt{n}\,|n-1\rangle \\ &= \sum_n \alpha C_n|n\rangle \end{aligned} \tag{5.80}$$

Performing an index shift, we find

$$C_{n+1}\sqrt{n+1} = \alpha C_n \tag{5.81}$$

which leads to

$$|\alpha\rangle = C_0\sum_n \frac{\alpha^n}{\sqrt{n!}}|n\rangle \tag{5.82}$$

The overall factor C_0 is determined by requiring that the coherent states be normalized

$$\begin{aligned} \langle\alpha|\alpha\rangle &= |C_0|^2\sum_{n,m}\frac{\alpha^n\alpha^{*m}}{\sqrt{n!m!}}\langle n|m\rangle \\ &= |C_0|^2\sum_n\frac{|\alpha|^{2n}}{n!} \\ &= |C_0|^2\, exp(|\alpha|^2) \\ &= 1 \end{aligned} \tag{5.83}$$

which gives us $C_0 = \exp(-|\alpha|^2/2)$ where we have absorbed an arbitrary phase and chosen C_0 to be real. The annihilation operator a is not Hermitian, so it is not guaranteed to have real eigenvalues, or orthogonal eigenstates. The coherent states are nonorthogonal states characterized by a complex number α.

What about fluctuations in the coherent state? If we define $\alpha = |\alpha|e^{i\phi}$, then we have

$$\langle X \rangle = |\alpha| \cos(\theta) \tag{5.84}$$

$$\langle Y \rangle = |\alpha| \sin(\theta) \tag{5.85}$$

$$\langle X^2 \rangle = \frac{1}{4} + |\alpha|^2 \cos^2(\theta) \tag{5.86}$$

$$\langle Y^2 \rangle = \frac{1}{4} + |\alpha|^2 \sin^2(\theta) \tag{5.87}$$

$$\langle N \rangle = |\alpha|^2 \tag{5.88}$$

$$\langle N^2 \rangle = |\alpha|^4 + |\alpha|^2 \tag{5.89}$$

$$\Delta N = |\alpha| \tag{5.90}$$

$$\Delta X = \frac{1}{2} \tag{5.91}$$

$$\Delta Y = \frac{1}{2} \tag{5.92}$$

There are a variety of interesting results here. The first is that the coherent state is a minimum uncertainty state satisfying $\Delta X \Delta Y = 1/4$. The second is that the fractional uncertainty in the photon number is

$$\frac{\Delta N}{\langle N \rangle} = \frac{1}{\sqrt{\alpha}} \tag{5.93}$$

Thus by increasing the value of $|\alpha|$ we can reduce the fractional uncertainty in amplitude to an arbitrarily small number. Also, just via the relation for arc length, we can write down $\Delta\phi \approx 1/(2|\alpha|) = 1/2\bar{n}$ which is only valid for $|\alpha| \gg 1$. As in our general discussion of a phase operator, we know that problems with definitions of a phase operator, and accompanying uncertainty relations, arise when the average field amplitude is small.

If a measurement of photon number is made on a coherent state, the probability of obtaining the result n is just

$$P_n = |C_n|^2 = e^{-|\alpha|^2/2} \frac{\alpha^n}{\sqrt{n!}} \tag{5.94}$$

which is a Poisson distribution. We could have surmised this, as for a coherent state the variance $(\Delta N)^2 = \langle N \rangle$, that is, the variance is equal to the mean, which is the hallmark of the Poisson distribution. But of course we need to look at the entire distribution as we have done to confirm a suspicion that the photon number distribution is Poissonian in nature. A fuzzball cartoon of the quadrature distribution of a coherent state is shown in figure 5.4. We see that the amplitude fluctuation is on the order of $\sqrt{\langle n \rangle} = \alpha$. There is a notion of phase, with a variance of $\Delta\phi = 1/|\alpha|$, coherent states are useful for interferometry, with a phase uncertainty that decreases the larger the amplitude $|\alpha|$, making interferometry possible with strong classical lasers. The quantum noise limit for a coherent state is a limiting factor in metrology, in particular gravitational wave detection.

In figure 5.5, we reproduce the cartoon from figure 5.4, but noting the quadrature uncertainties. Unlike the phase uncertainty which is heuristic, these are well defined quantum variances.

We have seen that at least formally, we can generate any field state by applying an operator onto the ground state, what is that operator for the coherent state? We have

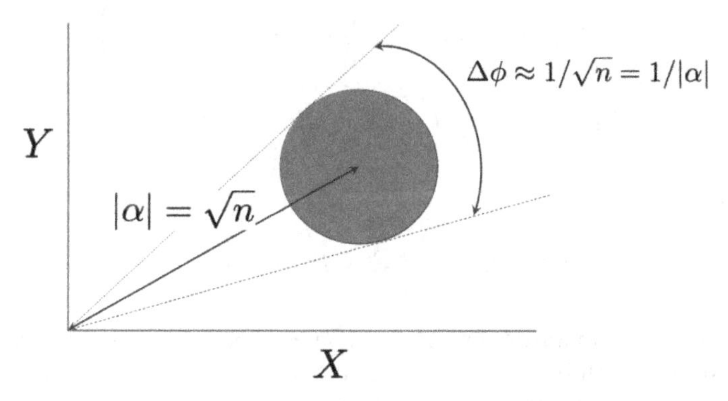

Figure 5.4. Cartoon of the quadrature distribution of a coherent state, noting phase and amplitude uncertainties.

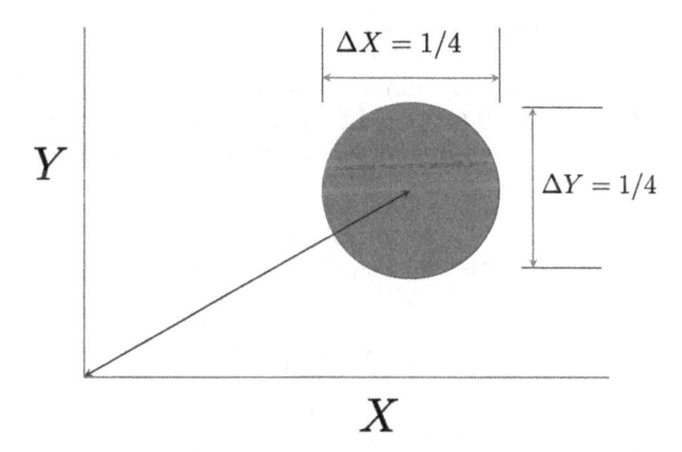

Figure 5.5. Cartoon of the quadrature distribution of a coherent state, noting X and Y quadrature uncertainties.

$$
\begin{aligned}
|\alpha\rangle &= \sum_n e^{-|\alpha|^2} \frac{\alpha a^{\dagger n}}{n!} |0\rangle \\
&= e^{-|\alpha|^2} e^{\alpha a^\dagger} |0\rangle
\end{aligned}
\tag{5.95}
$$

As the action of the lowering operator on the ground state yields zero, so does any function of the lowering operator. We are able to use $e^{-\alpha^* a}|0\rangle = |0\rangle$ which leads to

$$
\begin{aligned}
|\alpha\rangle &= e^{-|\alpha|^2} e^{\alpha a^\dagger} e^{-\alpha^* a} |0\rangle \\
&= e^{\alpha a^\dagger - \alpha^* a} |0\rangle \\
&= D(\alpha)|0\rangle
\end{aligned}
\tag{5.96}
$$

where we define the displacement operator $D(\alpha)$. We have had to be careful with commutation relations here, recall that for operators

$$
\begin{aligned}
e^A e^B &= \left(1 + A + \frac{1}{2}A^2 + \cdots\right)\left(1 + B + \frac{1}{2}B^2 + \cdots\right) \\
&= 1 + (A + B) + \frac{1}{2}(A^2 + B^2 + 2AB) + \cdots
\end{aligned}
\tag{5.97}
$$

while we also have

$$
e^{A+B} = 1 + (A + B) + \frac{1}{2}(A^2 + B^2 + AB + BA) + \cdots \tag{5.98}
$$

which shows that for operators $\exp(A)\exp(B) \neq \exp(A + B)$. If the commutator of A and B commutes with both operators, that is if $[A, [A, B]] = [B, [A, B]] = 0$, then a version of the Baker–Campbell–Hausdorff theorem yields

$$
e^A e^B = e^{A+B+\frac{1}{2}[A,B]} \tag{5.99}
$$

We have used this to write the displacement operator in the form

$$
D(\alpha) = e^{-|\alpha|^2} e^{\alpha a^\dagger} e^{-\alpha^* a} = e^{\alpha a^\dagger - \alpha^* a} \tag{5.100}
$$

As $D(\alpha)$ is an exponentiation of i times a Hermitian operator, it is then a unitary operator, $D^\dagger(\alpha)D(\alpha) = D(\alpha)D^\dagger(\alpha) = 1$. Let us write $D(\alpha)$ not in terms of creation and annihilation operators, but in terms of position and momentum operators, for real values of α,

$$
\begin{aligned}
D(\alpha) &= \exp\left(i\alpha\left(\frac{2}{\hbar\omega}\right)^{1/2} p\right) \\
&= \exp\left(-\alpha\left(\frac{2\hbar}{\omega}\right)\frac{d}{dx}\right) \\
&= \exp\left(x_\alpha \frac{d}{dx}\right)
\end{aligned}
\tag{5.101}
$$

where we have gone to the position representation and defined a displacement distance x_0. Consider now the form of a wavefunction $\psi(x)$ by some amount x_α, utilizing a Taylor series expansion,

$$\begin{aligned}\psi &= \psi(x) + x_\alpha \frac{d\psi}{dx}\bigg|_x + \frac{1}{2}x_\alpha^2 \frac{d^2\psi}{dx^2}\bigg|_x + \cdots \\ &= \exp\left(x_\alpha \frac{d}{dx}\right)\psi(x)\end{aligned} \tag{5.102}$$

where we have summed the series. So, it seems that the displacement operator does exactly that, it picks up the wave function and moves it over to a new position. The state that is displaced in this case to form a coherent state is the ground, or vacuum state. Thus, we can see the fact that the 'noise' in the coherent state is the same as the vacuum state can be more easily explained.

We can also see how the displacement operator acts on operators,

$$D^\dagger(\alpha)aD(\alpha) = a + \alpha \tag{5.103}$$

$$D^\dagger(\alpha)a^\dagger D(\alpha) = a^\dagger + \alpha^* \tag{5.104}$$

Similarly, we have

$$D^\dagger(\alpha)XD(\alpha) = X + \Re(\alpha) \tag{5.105}$$

$$D^\dagger(\alpha)YD(\alpha) = Y + \Im(\alpha) \tag{5.106}$$

These relations, when utilized when acting on the vacuum state, yield all the formulas for the averages and fluctuations we have derived earlier. For example,

$$\begin{aligned}\langle\alpha|N|\alpha\rangle &= \langle 0|D^\dagger(\alpha)ND(\alpha)|0\rangle \\ &= \langle 0|D^\dagger(\alpha)a^\dagger D(\alpha)D^\dagger(\alpha)aD(\alpha)|0\rangle \\ &= \langle 0||\alpha|^2 + \alpha a^\dagger + \alpha^* a + a^\dagger a|0\rangle \\ &= |\alpha|^2\end{aligned} \tag{5.107}$$

What about the time dependence of the coherent state? The time dependence of a number state is just $|n(t)\rangle = \exp(-iHt/\hbar)|n(0)\rangle = \exp(-i(n + 1/2)\omega t)|n(0)\rangle$. The time dependence of the coherent state is then

$$\begin{aligned}|\alpha(t)\rangle &= e^{-i\omega t/2}\sum_n \frac{\alpha^n e^{-in\omega t}}{\sqrt{n!}}|n(0)\rangle \\ &= \sum_n \frac{(\alpha e^{-i\omega t})^n}{\sqrt{n!}}|0\rangle \\ &= \sum_n \frac{\alpha(t)}{\sqrt{n!}}|0\rangle\end{aligned} \tag{5.108}$$

where we have suppressed a time dependent but overall phase. Thus we see that the time dependence of the coherent state is merely the replacement of $\alpha(t) = \alpha(0)e^{-i\omega t}$, a coherent state is ground state quantum noise superimposed over a classical-like phasor. In terms of a mechanical oscillator in a quadratic potential well, the coherent state is a Gaussian of ground state width that merely oscillates back and forth in the well, without spreading. In many ways, the coherent state is the most classical-like state of the oscillator, where the classical mass oscillating as $x = x_0 \cos(\omega t + \phi)$ is replaced by our quantum 'fuzzball' (or Gaussian wave function).

The time dependence of a coherent state in terms of a wave function oscillating in a well is shown in figure 5.6. We have a Gaussian wave function that is displaced from the origin, and which maintains its shape as it propagates, much like a classical mass on a spring. How does this arise? We know that states of different energy should have different velocities, indeed this leads to dispersion of a pulse during propagation in free space. In the case of the quadratic potential, the higher energy parts of the wave travel faster, but have a larger amplitude of oscillation. The two factors balance one another, and a Gaussian wavefunction will maintain its shape as it oscillates. In terms of quadrature amplitudes, we have a Gaussian fuzzball that rotates in the $X-Y$ plane at frequency ω.

The photon number distribution for a coherent state is a Poissonian distribution, with mean equal to the variance, as shown in figure 5.7.

The noise, or variance, in a coherent state remains constant as it evolves in time; recall that it remains a coherent state with a different complex phase. In figure 5.8, we show the time evolution of the photon number (constant) and the quadrature amplitudes X and Y.

It is not terribly convenient to use the coherent states as a basis, they are not orthonormal,

$$\langle\alpha|\alpha'\rangle = e^{\frac{-1}{2}|\alpha-\alpha'|^2} \tag{5.109}$$

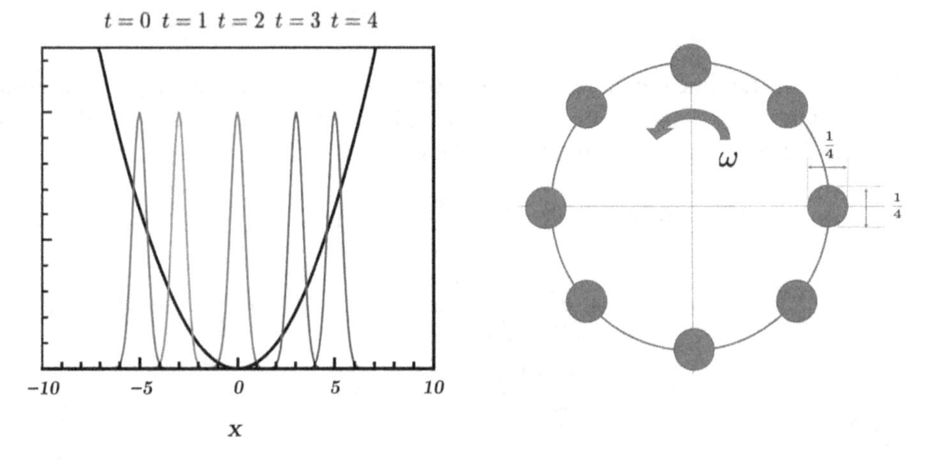

Figure 5.6. Time evolution of a coherent state wavefunction, and quadrature distribution cartoon.

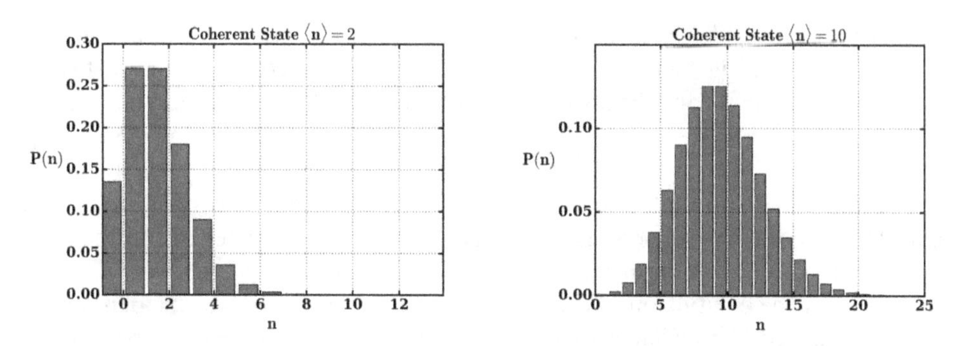

Figure 5.7. Photon probability distribution $P(n)$ for coherent states with average photon numbers 2 and 10.

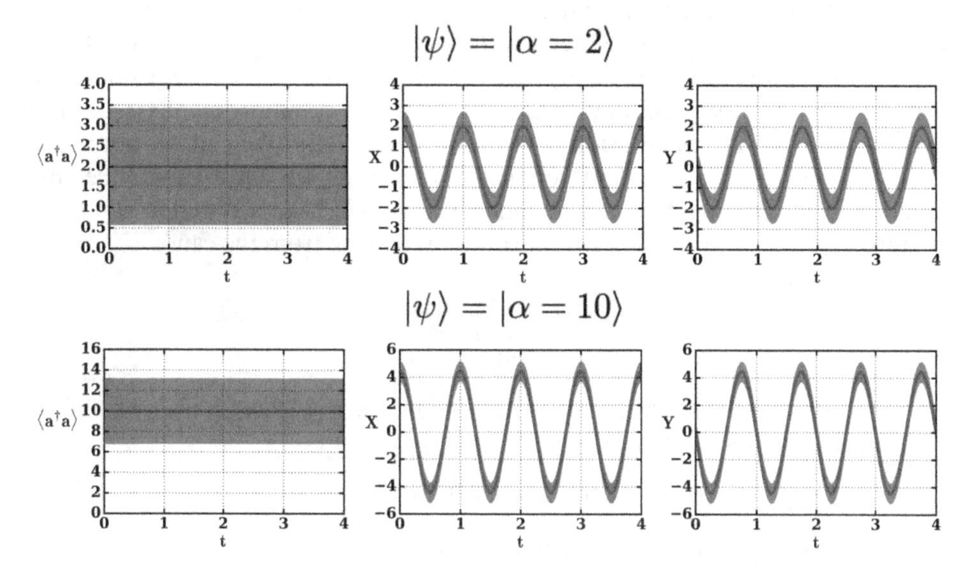

Figure 5.8. Time evolution of the average (blue curve) and variance (red shaded region) of coherent states with average photon numbers of 2 and 10, respectively.

and indeed form an overcomplete set. They can resolve the identity,

$$I = \frac{1}{\pi}\int_{-\infty}^{\infty} |\alpha\rangle\langle\alpha| d^2\alpha \tag{5.110}$$

How does one generate a coherent state in the lab? Well if we have a Hamiltonian of the form

$$H = i\hbar(\beta a^{\dagger} - \beta^{*} a) \tag{5.111}$$

then the time evolution of the vacuum for some time T would yield

$$\psi(t) = D(\beta T)\psi(0) \tag{5.112}$$

which would be a coherent state with $\alpha = \beta T$. This Hamiltonian can come from a linear coupling of two field modes, such as one field incident on an imperfect end mirror of a cavity. The interaction time would be of the order of $1/\kappa$ and the field inside the cavity would reach a steady state described by a coherent state. Of course the real Hamiltonian for a linear coupling between field modes would be

$$H = i\hbar(ba^{\dagger} - b^{\dagger}a) \tag{5.113}$$

where b and $b^{\dagger}$ are annihilation and creation operators for the driving mode. This kind of interaction kills a photon in the driving mode and creates one in the cavity being driven. The other term represents the reverse process, which would be light leaking out of the cavity from the same mirror the driving field was incident on. We regain the needed Hamiltonian to generate a coherent state if we can replace $b \rightarrow \beta$ and $b^{\dagger} \rightarrow \beta^{*}$. This would be appropriate for a strong coherent state, but we are trying to MAKE a coherent state, so this becomes circular! A strong steady oscillating current will generate a coherent state to a good approximation, but we will see in our work on laser theory that a laser well above threshold produces a coherent state to very good approximation, with the addition of phase diffusion noise due to spontaneous emission noise in the laser. So, in the end, we get a laser, run it well above threshold, and conduct our experiments on a time scale small compared to the phase diffusion rate, and then we will have a coherent state.

Another way to generate a coherent state can be seen from the canonical form of the atom–field interaction,

$$H_{\text{int}} = -\frac{e}{m}\vec{A} \cdot \vec{p} \tag{5.114}$$

The current density for charges moving at velocity v is $\vec{J} = e\vec{v} = (e/m)\vec{p}$, which leads to an interaction Hamiltonian

$$H_{\text{int}} = -\vec{J} \cdot \vec{A} \tag{5.115}$$

Recall the form of the quantized vector potential

$$A = A_0(ae^{-i\omega t} + a^{\dagger}e^{i\omega t}) \tag{5.116}$$

If we have an oscillatory current density in one dimension $J = J_0 \cos(\omega t)$, then we are led to a time evolution operator

$$U(t) = e^{-iA_0J_0(a+a^{\dagger})t} \tag{5.117}$$

which is of the form of a displacement operator that will generate a coherent state.

5.6 Squeezed states

The coherent states are minimum uncertainty states, that have symmetric noise, in the sense that any quadrature $a_{\theta} = (1/2)(ae^{i\theta} + a^{\dagger}e^{-i\theta})$ satisfies $\Delta a_{\theta} = 1/2$. Here, we wish to consider a class of states that are minimum uncertainty states, but do not have symmetric noise. We have number squeezed states and phase squeezed states, these could be more useful for measurements of amplitude and phase than coherent states as we shall see. States of this type are not generated in a linear interaction, as we saw that would produce a coherent state. Let us consider an analogy of the displacement operator, but one that involves a^2 and $a^{\dagger 2}$, the squeezing operator

$$S(r) = \exp\left(\frac{1}{2}(ra^2 - r^* a^{\dagger 2})\right) \tag{5.118}$$

Let us examine the action of this operator on a vacuum state, which generates what is known as a squeezed vacuum state. One can use a Taylor series approximation of the squeezing operator, and take r to be real for now, to order α^2,

$$S(r) = 1 + \frac{r}{2}(a^2 - a^{\dagger 2}) + \frac{r^2}{4}(a^4 + a^{\dagger 4} - a^2 a^{\dagger 2} - a^{\dagger 2} a^2) + \cdots \tag{5.119}$$

and generate

$$S(r)|0\rangle = |0\rangle - \frac{1}{\sqrt{2}} r|2\rangle + \frac{r^2}{2}|0\rangle + 6r^2|2\rangle + \cdots \tag{5.120}$$

Here, we see that due to the $a^{\dagger 2}$ terms, we find photons in pairs in the sense that only even numbered number states show up in the squeezed vacuum state. This points to the fragility of squeezed vacuum to loss, a coherent state is of the form (to order α^2)

$$|\alpha\rangle \approx (1 - |\alpha|^2)|0\rangle + \alpha|1\rangle + \frac{\alpha^2}{\sqrt{2}}|2\rangle \tag{5.121}$$

and so losing photons has the potential to turn a squeezed vacuum into something more like a coherent state, and the noise will be symmetrized. Using the Baker–Hausdorff theorem, we can show

$$b = S^\dagger(r) a^\dagger S(r) = a^\dagger \cosh(|r|) - a e^{-i\phi} \sinh(|r|) \tag{5.122}$$

$$b^\dagger = S^\dagger(r) a S(r) = a \cosh(|r|) - a^\dagger e^{-i\phi} \sinh(|r|) \tag{5.123}$$

where we have written the squeezing parameter $r = |r|\ exp^{(i\phi)}$.

This is known as a Bogoliubov transformation and can be written in matrix form as

$$\begin{pmatrix} b \\ b^\dagger \end{pmatrix} = \begin{pmatrix} \cosh(r) & -\sinh(r) \\ -\sinh(r) & \cosh(r) \end{pmatrix} \begin{pmatrix} a \\ a^\dagger \end{pmatrix} \tag{5.124}$$

The quadratures also transform under the same transition

$$\begin{pmatrix} X' \\ Y' \end{pmatrix} = \begin{pmatrix} \cosh(r) & -\sinh(r) \\ -\sinh(r) & \cosh(r) \end{pmatrix} \begin{pmatrix} X \\ Y \end{pmatrix} \tag{5.125}$$

You have perhaps seen a transformation similar to this before in special relativity

$$\begin{pmatrix} x' \\ ct' \end{pmatrix} = \begin{pmatrix} \gamma & -\gamma v/c \\ -\gamma v/c & \gamma \end{pmatrix} \begin{pmatrix} x \\ ct \end{pmatrix} \tag{5.126}$$

with

$$\gamma = \sqrt{1 - v^2/c^2} \tag{5.127}$$

known as the Lorentz transformation. The Boglioubov transformation is a Lorentz transformation with $\tanh(r) = v/c$ playing the role of the rapidity. The creation and annihilation operators get mixed, and hence so do absorption and emission. This will inform our later discussions of how an accelerated detector can be excited, the Unruh effect, as well as our brief discussion of Hawking radiation.

In general, we can expand the squeezed vacuum in a superposition of number states,

$$|r\rangle = S(r)|0\rangle = \sum_n D_n|n\rangle \tag{5.128}$$

It will be useful to consider

$$\begin{aligned} 0 &= a|0\rangle \\ &= S(r)a|0\rangle \\ &= S(r)aS^\dagger(r)S(r)|0\rangle \\ &= S(r)aS^\dagger(r)|r\rangle \\ &= (a\cosh(|r|) - a^\dagger e^{-i\phi}\sinh(|r|))|r\rangle \end{aligned} \tag{5.129}$$

Then we can write

$$\begin{aligned} 0 &= a\cosh(|r|) + a^\dagger e^{-i\phi}\sinh(|r|)\sum_n D_n|n\rangle \\ &= \sum_n \cosh(|r|)D_n\sqrt{n}\,|n-1\rangle + \sum_n \sinh(|r|)e^{-i\phi}\sqrt{n+1}\,|n+1\rangle \\ &= \sum_n (\cosh(|r|)D_{n+1} + \sinh(|r|)e^{-i\phi}D_{n-1}) \end{aligned} \tag{5.130}$$

which leads to a recursion relation of the form

$$D_{n+1} = -e^{i\phi}\tanh(|r|)\sqrt{\frac{n}{n+1}}D_{n-1} \tag{5.131}$$

with D_0 to be determined from normalization and $D_1 = 0$. By requiring $\langle r|r\rangle = 1$ and the above recursion relation, we find

$$|r\rangle = \sqrt{|r|}\sum_n \frac{\sqrt{(2n)!}}{n!}\left(\frac{-\tanh(|r|)e^{i\phi}}{2}\right)^n |2n\rangle \tag{5.132}$$

The squeezed vacuum state indeed contains only even numbered Fock states as we anticipated. These states are also called two-photon coherent states; squeezed vacuum is a two-photon coherent state, albeit with a zero coherent amplitude. We will consider squeezed states *with* a coherent amplitude shortly. For now we will consider r to be real. For the squeezed vacuum, we have

$$\langle X \rangle = 0 \tag{5.133}$$

$$\langle Y \rangle = 0 \tag{5.134}$$

$$\langle N \rangle = \sinh^2(r) \tag{5.135}$$

$$\langle N^2 \rangle = 3\sinh^4(r) + 2\sinh^2(r) \tag{5.136}$$

$$\Delta N = \sqrt{2\langle N \rangle(\langle N \rangle + 1)} \tag{5.137}$$

Squeezed vacuum can be a misnomer, it is generated by the squeezing operator on the vacuum, but we see that there is indeed a nonzero photon number. The quadrature variances are given by

$$(\Delta X)^2 = \frac{1}{4}e^{-2r} \tag{5.138}$$

$$(\Delta Y)^2 = \frac{1}{4}e^{2r} \tag{5.139}$$

with

$$\Delta X \Delta Y = \frac{1}{4} \tag{5.140}$$

which shows that the squeezed vacuum is a minimum uncertainty state with asymmetric noise. In general, if the squeezing parameter is complex (nonzero ϕ, meaning the squeeze does not occur along the X-axis), the variances are mixed, and the 'quiet' quadrature is not X, but the quadrature defined by a_ϕ. The variances for X and Y are

$$(\Delta X)^2 = \frac{1}{4}(e^{-2|r|}\cos^2(\phi/2) + e^{2|r|}\sin^2(\phi/2)) \tag{5.141}$$

$$(\Delta Y)^2 = \frac{1}{4}(e^{2|r|}\cos^2(\phi/2) + e^{-2|r|}\sin^2(\phi/2)) \tag{5.142}$$

This is shown in figure 5.9.

The uncertainty principle $\Delta X \Delta Y \geqslant 1/4$ limits the available physical states. This is shown in figure 5.10. Here, all the coherent states lie at the point $\Delta X = \Delta Y = 1/4$. For the squeezed vacuum one can have less noise in one quadrature, below the coherent state limit of 1/4 with a corresponding increase of noise in the other quadrature. States that live between the red and blue curves are all referred to as squeezed states, as one quadrature has noise $\Delta X(Y) \leqslant 1/4$. The squeezed vacuum states lie on the blue curve. These types of states were first observed by Slusher *et al* [9].

In figure 5.11, we show a sketch of the quadrature distribution for a squeezed state created via displacement of a squeezed vacuum state.

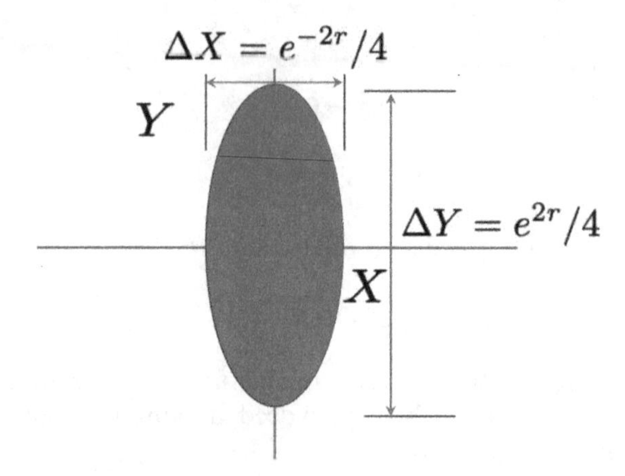

Figure 5.9. Cartoon of the quadrature distribution of a squeezed state, noting X and Y quadrature uncertainties.

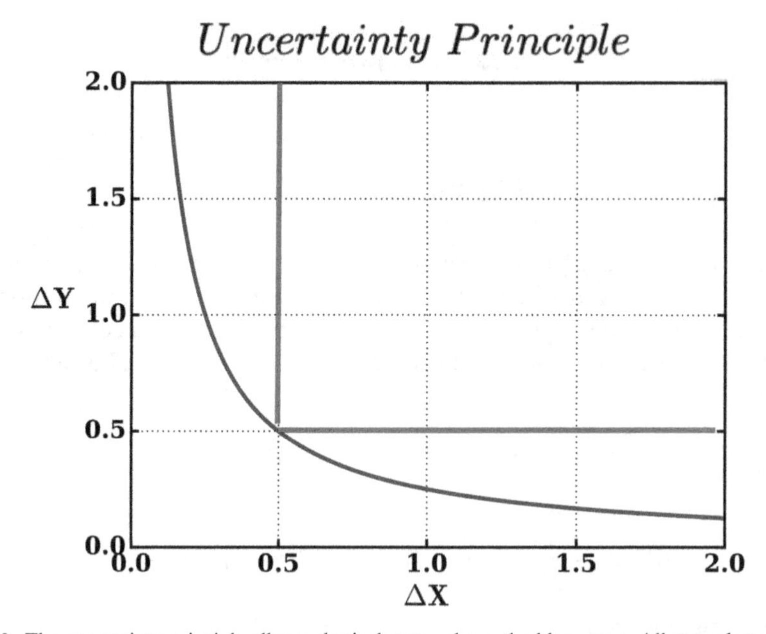

Figure 5.10. The uncertainty principle allows physical states above the blue curve. All states between the red and blue curves are referred to as squeezed states. Squeezed vacuum states lie on the blue curve.

The actions of squeezing and displacement do not commute, but they can be related

$$D(\alpha)S(r) = S(r)D(\beta) \tag{5.143}$$

with the displacements α and β related

$$\beta = \alpha \cosh(r) + \alpha^* \sinh(r) \tag{5.144}$$

essentially a Bogoloubov transformation of the displacements.

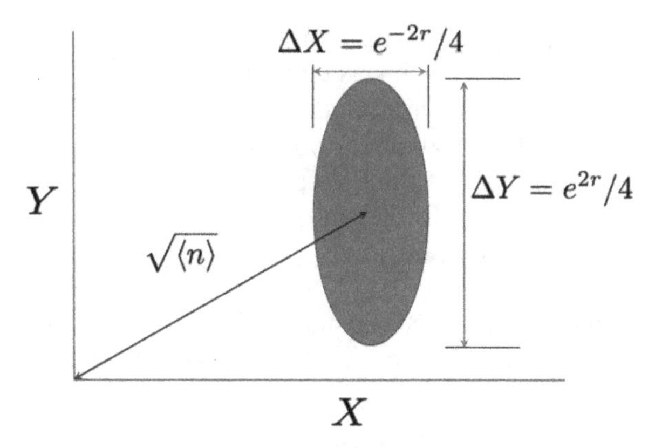

Figure 5.11. Cartoon of the quadrature distribution of a squeezed state, noting X and Y quadrature uncertainties.

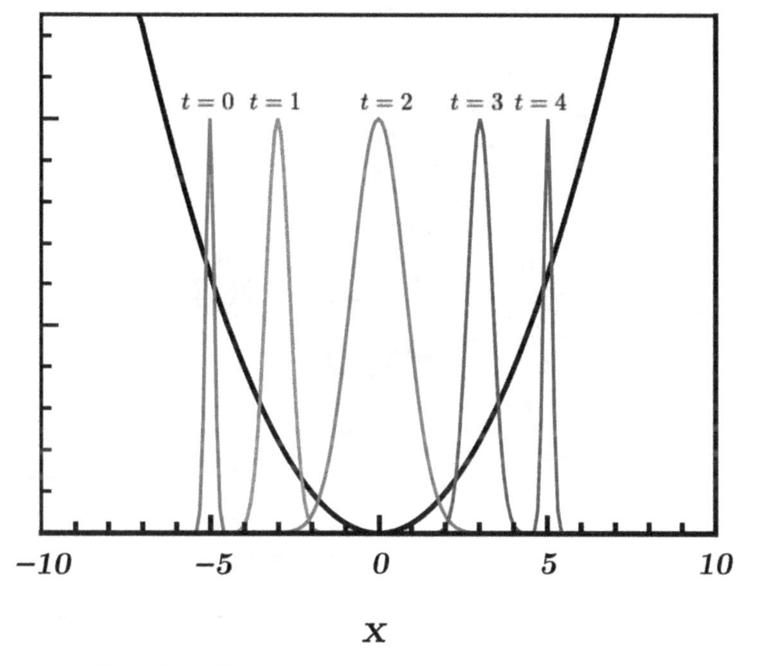

Figure 5.12. Time evolution of a squeezed state wave function.

We have cheated a bit so far, we have characterized the various quadratures, and have talked primarily about phases $\theta = 0$ and $\pi/2$. Well as the phase of the field tends to have a contribution ωt, then the various properties of the squeezed vacuum are time dependent. At $t = \pi/2\omega$, the quiet phase that has variance $e^{-2r}/4$ at $t = 0$ will be the noisy phase. So, to always 'see' the quiet phase, we must just look at every half cycle; there are measurement schemes that do precisely that. This is shown in figure 5.12. In terms of a wave packet in a quadratic potential well, at $t = 0$ if the wave packet is at the origin and is narrower than the coherent state width, at a quarter of a cycle, the wave packet will be larger than a coherent state width. A half cycle later it is back at the origin as it was initially, and then it repeats this motion to the other side of the well, as shown in figure 5.12.

The time evolution of this fuzzball cartoon is shown in figure 5.13.

The squeezed vacuum only has nonzero occupation probability for even photon numbers. As the squeezing operator depends on a^2, applying that to a vacuum state to generate the squeezed vacuum only even photon numbers obtain. Two examples are shown in figure 5.14.

How does one generate squeezed vacuum? The initial state is trivial, being the vacuum state. The squeezing operator can be generated by a Hamiltonian

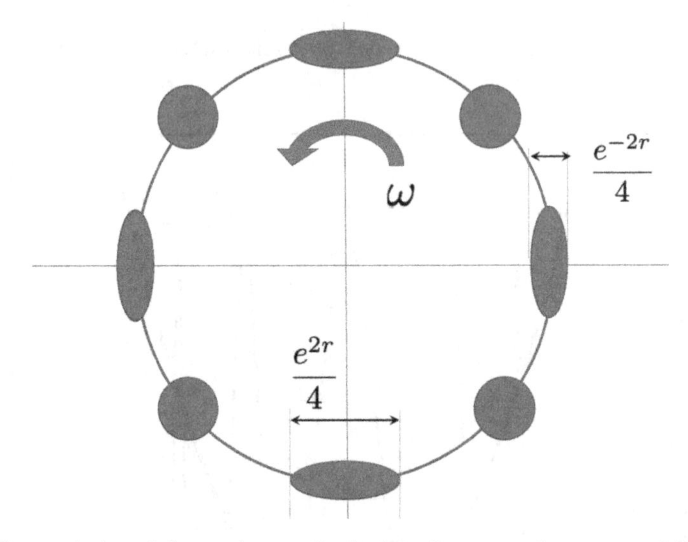

Figure 5.13. Time evolution of the quadrature distribution of a squeezed state, essentially a quadrapole oscillation.

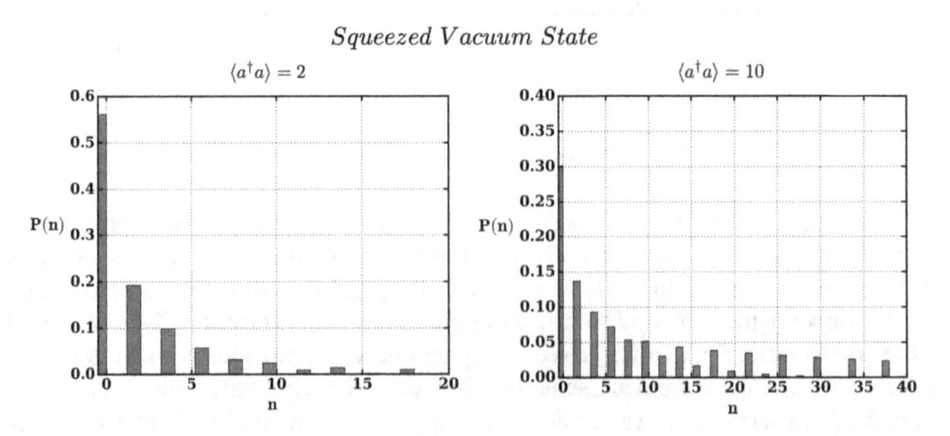

Figure 5.14. Photon probability distribution $P(n)$ for squeezed vacuum states with average photon numbers 2 and 10.

$$H = \frac{i\hbar}{2}(sa^2 - s^* a^{\dagger 2}) \tag{5.145}$$

which would lead to the state

$$\psi(t) = e^{\frac{1}{2}(sa^2 - s^* a^{\dagger 2})T}\psi(0) \tag{5.146}$$

after time evolution for $t = T$, which is a squeezed state with $r = sT$. What physical process would correspond to such a Hamiltonian? Well one that creates or destroys two photons at a time (figure 5.15). What if we have a material that has a nonlinear polarization $P = \varepsilon_0 \chi^{(2)} E^2$, while not pretending to do a nonlinear version of a quantized field, this would look something like

$$P = \varepsilon_0 \chi^{(2)} \mathcal{E}_0^2 (a^2 + a^{\dagger 2} + a^\dagger a + aa^\dagger) \tag{5.147}$$

This looks promising, and indeed a material that has a second-order nonlinearity can be used to generate squeezed vacuum. These types of materials will take a laser of one frequency ω, absorbing two photons, and then output one photon at twice the frequency 2ω, a process known as second harmonic generation. This process results from the $a^{\dagger 2}$ term in the Hamiltonian, where s would be proportional to the incident field amplitude at the fundamental frequency. The opposite process is parametric down conversion, where a material with a second-order nonlinearity will absorb a photon at ω and emit two photons at frequencies ω_1 and ω_2, that due to energy conservation need to satisfy $\omega = \omega_1 + \omega_2$. If this is done with a nonlinear crystal inside a cavity, we can have a system that reaches a steady state, where the gain of the parametric process equals the loss through the end mirrors, just as in a laser. This is generally known as parametric oscillation. If $\omega_1 = \omega_2 = \omega/2$, that is known as degenerate parametric oscillation. So, to generate squeezed vacuum, we pump a

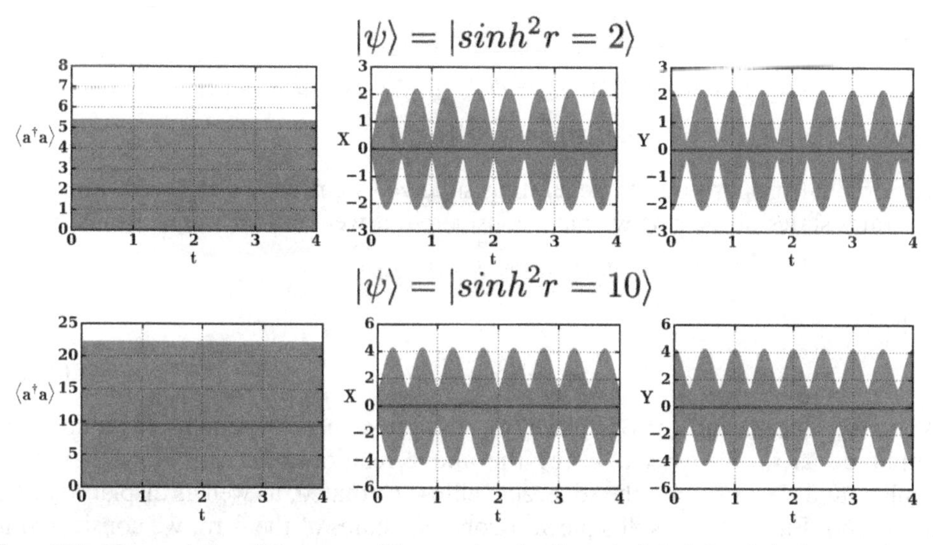

Figure 5.15. Time evolution of the average (blue curve) and variance (red shaded region) of squeezed states with average photon numbers of 2 and 10, respectively.

nonlinear system in a cavity with light at 2ω, and light at ω will spontaneously arise at frequency ω (assuming the cavity has a mode at ω), which is in a squeezed vacuum state. Of course there are technical problems that arise that we are completely ignoring here. We will discuss how to detect squeezed vacuum, and how to interrogate quantum 'fuzzballs' in chapter 6.

What if we take our optical parametric oscillator, and input a coherent state of light at frequency ω, we would then generate a state

$$|SS\rangle = S(r)D(\alpha)|0\rangle \tag{5.148}$$

These states will be minimum uncertainty states with asymmetric noise, and also have a coherent amplitude. Mathematically, we get states that are easier to deal with if we reverse the order of the operators, to generate the ideal squeezed coherent states,

$$|r, \alpha\rangle = D(\alpha)S(r)|0\rangle \tag{5.149}$$

they are a squeezed vacuum that is then displaced. They are different from the states generated by squeezing a coherent state, as the squeezing operator and the displacement operator do not commute. Again, we examine the effect of transforming operators instead of states.

$$D^{\dagger}(\alpha)S^{\dagger}(r)a^{\dagger}S(r)D(\alpha) = a^{\dagger}\cosh(|r|) - ae^{-i\phi}\sinh(|r|) + \alpha \tag{5.150}$$

$$D^{\dagger}(\alpha)S^{\dagger}(r)aS(r)D(\alpha) = a\cosh(|r|) - a^{\dagger}e^{-i\phi}\sinh(|r|) + \alpha^{*} \tag{5.151}$$

The variances for X and Y are the same as for a squeezed vacuum, as they are merely displaced. The average photon number is

$$\langle N\rangle = \sinh^2(|r|) + |\alpha|^2 \tag{5.152}$$

with variance

$$\begin{aligned}(\Delta N)^2 = {}&|\alpha|^2\left(e^{2|r|}\sin^2(\theta - \phi/2) + e^{-2|r|}\cos^2(\theta - \phi/2)\right)\\ &+ 2\sinh^2(|r|)(\sinh^2(|r|) + 1)\end{aligned} \tag{5.153}$$

Our original proposition for generating squeezed light with a coherent amplitude generates states of the sort we have considered, if we consider the relations

$$D(\alpha)S(r) = S(r)D(\alpha') \tag{5.154}$$

with

$$\alpha' = \alpha r + r\alpha \tag{5.155}$$

When we squeeze and then displace, we just get squeezed vacuum plus a coherent amplitude as we have seen. Using the above relations, we can show that if we displace and then squeeze, the squeezing ellipse is rotated, as well as displaced along an arc too. But they are still squeezed coherent states of the form we considered in detail, albeit with different parameters,

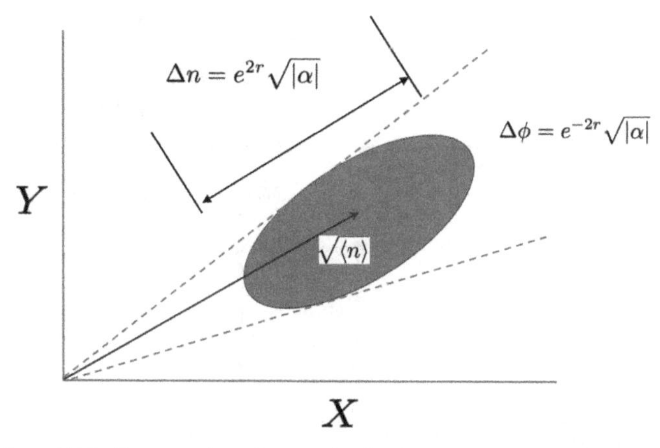

Figure 5.16. Cartoon of quadrature distribution for a phase squeezed state.

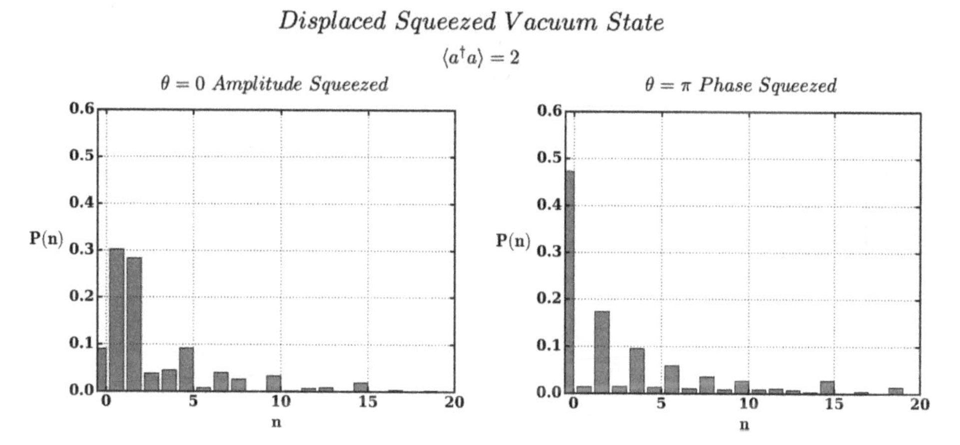

Figure 5.17. Photon number distribution for an average photon number of 2, $\sinh^2(r) = 1$ and $|\alpha|^2 = 1$. Shown are $\theta = 0$ and $\theta = \pi$.

$$S(r)D(\alpha)|0\rangle = D(\alpha')S(r)|0\rangle = |r, \alpha'\rangle \tag{5.156}$$

More schematically, we can talk about phase and amplitude squeezed states, in which the squeezing is parallel or perpendicular to the coherent amplitude; $\theta = \phi/2$ for phase squeezing and $\theta - \phi/2 = \pi/2$ for amplitude squeezed light (figure 5.16). This is shown in figure 5.17 for a displaced squeezed vacuum state, with an average photon number of 2, $\sinh^2(r) = 1$ and $|\alpha|^2 = 1$. In figure 5.18, we have an average photon number of 10, $\sinh^2(r) = 5$ and $|\alpha|^2 = 5$.

5.7 Cat states

Coherent states are essentially classical states with added vacuum noise. Recall the Schrödinger cat thought experiment where

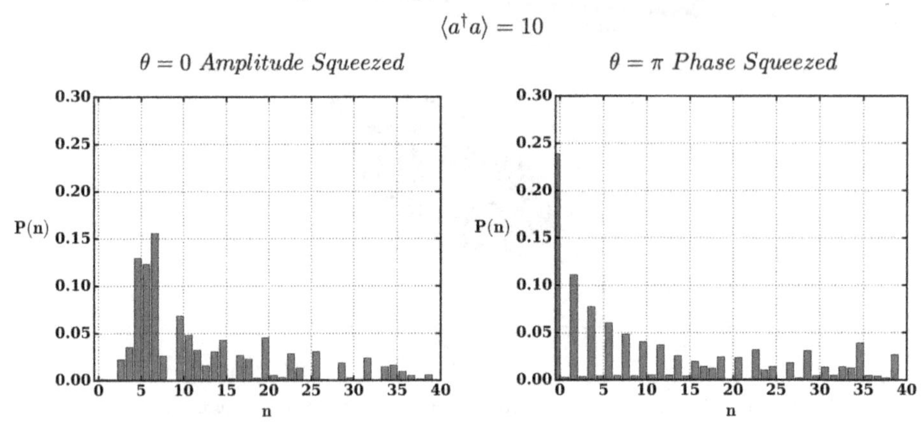

Figure 5.18. Photon number distribution for an average photon number of 10, $\sinh^2(r) = 5$ and $|\alpha|^2 = 5$. Shown are $\theta = 0$ and $\theta = \pi$.

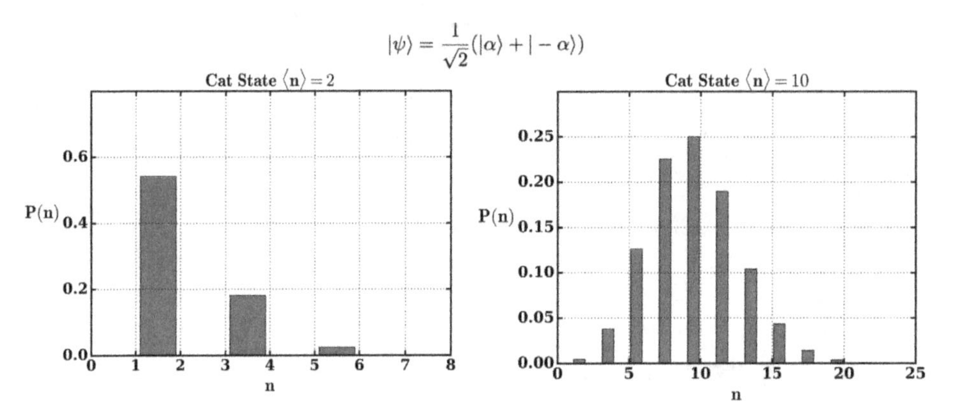

Figure 5.19. Photon number distributions for cat states with average photon numbers of 2 and 10.

$$|\Psi\rangle = \frac{1}{\sqrt{2}}(|\text{live}\rangle + |\text{dead}\rangle) \tag{5.157}$$

representing a cat that is, in some sense, alive and dead at the same time. The state of the cat is not known until it is measured. We remark that it is very hard to isolate quantum states from their environments. For a real box, if death is triggered by a radioisotope decay inside the box, within a lifetime or two the cat is dead. And for a real box, one would eventually notice the cat is dead due to a strong odor.

It turns out, we can superimpose two classical-ish states, a coherent state $|\alpha\rangle$ and one of similar amplitude but π out of phase (displaced in a negative direction) $|\alpha\rangle$. In this case, we know that the photon number distribution is Poissonian for each, but what is the result when they are superposed? In figure 5.19, we show the resulting photon distribution for the cat state

$$|Cat\rangle = \frac{1}{\sqrt{2}}(|\alpha\rangle + |-\alpha\rangle) \tag{5.158}$$

In figure 5.20, we show the time evolution of expectation values and variance, of n, X, and Y for cat states with average photon numbers of 2 and 10.

We see behavior like that of a squeezed vacuum state, where only even photon numbers have nonzero probability. This makes us suspect that this state is non-classical. It is, and when we discuss quasiprobability distributions in chapter 10, we can make that quantitative. Do we see something different with the state

$$|Cat\rangle = \frac{1}{\sqrt{2}}(|l\alpha\rangle - |-\alpha\rangle) \tag{5.159}$$

Not by simple quadrature and photon number measurements. We would need some type of interferometric measurement.

5.8 Thermal states

Finally, we examine our old friend the thermal state. A thermal state has no wave function, but is described by a density matrix. The diagonal elements of ρ_{thermal} are given by the Planck distribution, the off-diagonal elements are all zero. There is no coherence in a thermal state. The population distribution of a thermal state indeed decays exponentially, as displayed in figure 5.21.

The complete lack of any phase coherence in a thermal state can be seen in a plot of the expectation values and variances of the photon number and the X and Y quadratures. Nothing 'wavelike' is manifest. Examples are presented in figure 5.22.

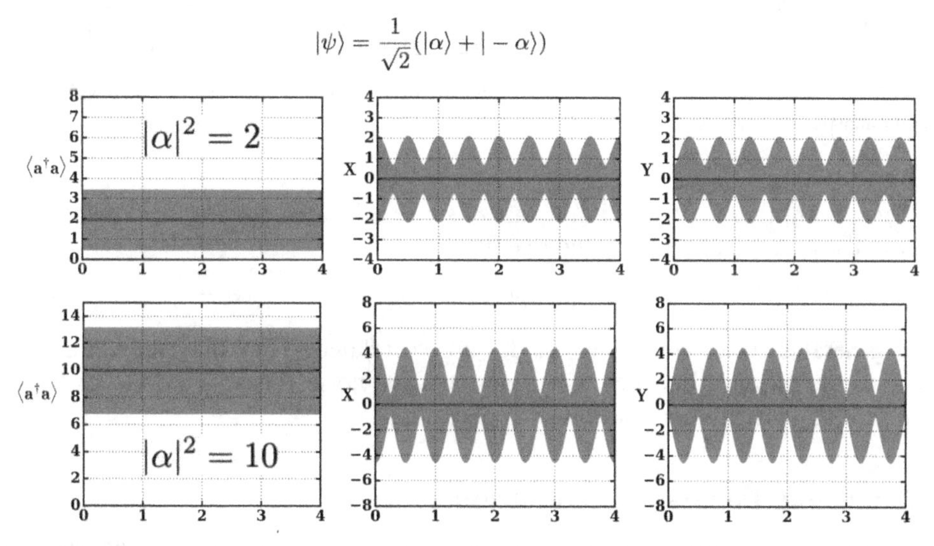

Figure 5.20. Time evolution of expectation values and variance, of n, X, and Y for cat states with average photon numbers of 2 and 10.

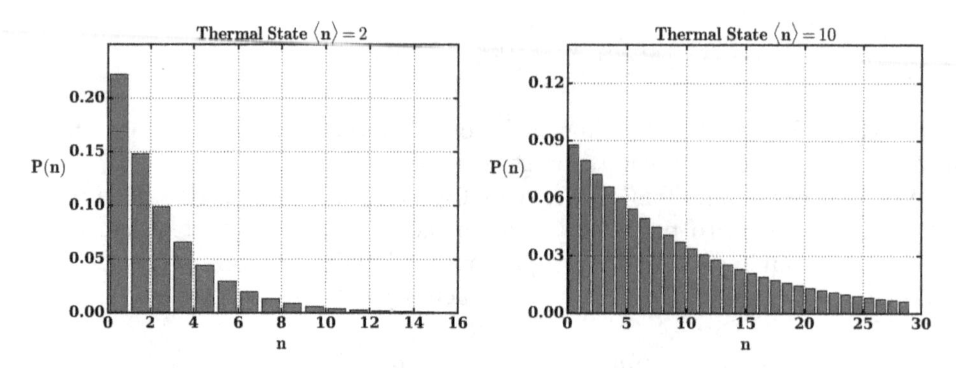

Figure 5.21. Photon number distribution for thermal states of average photon numbers of 2 and 10.

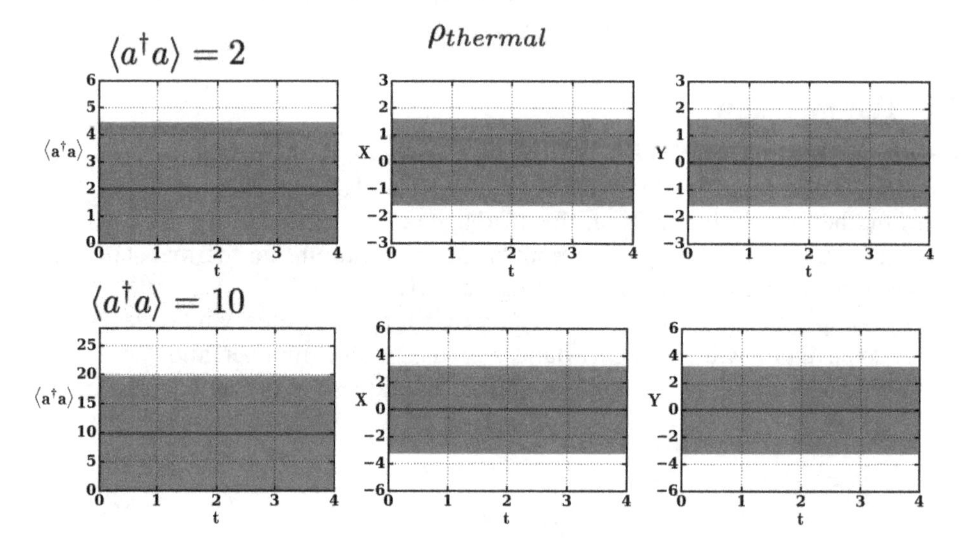

Figure 5.22. Time evolution of expectation values and variance, of n, X, and Y for thermal states with average photon numbers of 2 and 10.

There is another very important class of squeezed states, the so-called twin beams states. These are generated by the operator

$$T = e^{\eta(ab+a^{\dagger}b^{\dagger})} \tag{5.160}$$

This generates states where two modes are correlated. They may individually be noisy as the dickens, but they will have the *same* noise. These are quite useful for quantum key distribution and other applications.

5.9 Vacuum fluctuations and beam splitters

Consider the beam splitter shown in figure 5.23, with fields E_1 and E_2 incident, and fields E_3 and E_4 transmitted.

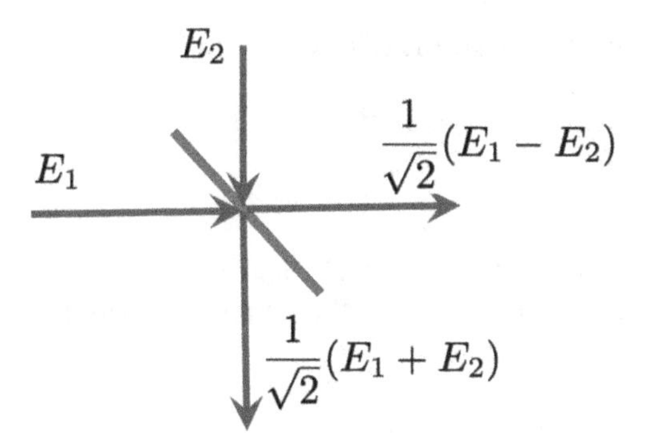

Figure 5.23. Inputs and outputs of a beam splitter.

These satisfy the relations

$$E_3 = rE_1 + tE_2 \tag{5.161}$$

$$E_4 = t'E_1 + r'E_2 \tag{5.162}$$

As the intensities are proportional to the square of the field, $I = \varepsilon_0 c|E|^2$, these are

$$I_3 = |r|^2 I_1 + |t|^2 I_2 + 2rt\varepsilon_0 cE_1E_2 \tag{5.163}$$

$$I_4 = |r'|^2 I_2 + |t'|^2 I_1 + 2r't'\varepsilon_0 cE_1E_2 \tag{5.164}$$

For energy to be conserved, we must have the relations

$$|r|^2 + |t|^2 = 1 \tag{5.165}$$

$$|r'|^2 + |t'|^2 = 1 \tag{5.166}$$

$$rt + r't' = 0 \tag{5.167}$$

These can be satisfied by taking $r = r'$ and $t' = -t$, and we see that there must be a relative π phase shift between the two transmitted parts. Now consider the quantum mechanical version, instead of fields we will use operators. We have modes described by a and b incident on the beamsplitter, with modes c and d in the outputs. We then have

$$c = ra + tb \tag{5.168}$$

$$d = -ta + rb \tag{5.169}$$

We can write this in matrix form as

$$\begin{pmatrix} c \\ d \end{pmatrix} = \begin{pmatrix} r & t \\ -t & r \end{pmatrix}\begin{pmatrix} a \\ b \end{pmatrix} \tag{5.170}$$

Let us consider the commutation relations for c and $c^\dagger$,

$$\begin{aligned}[c, c^\dagger] &= |r|^2[a, a^\dagger] + |t|^2[b, b^\dagger] + 2rt^*[a, b^\dagger] + 2r^*t[a^\dagger, b] \\ &= |r|^2 + |t|^2 \\ &= 1\end{aligned} \tag{5.171}$$

Similarly, $[d, d^\dagger] = 1$. What if we have no field input on the beam splitter in the b mode? Can we just set $b = 0$? If we do that, then we would have the commutation relations

$$[c, c^\dagger] = |r|^2 \neq 1 \tag{5.172}$$

$$[d, d^\dagger] = |t|^2 \neq 1 \tag{5.173}$$

which clearly would be a problem. We are reminded once again that there can be no absence of electric field quantum mechanically! The average value of the field $\langle a \rangle = 0$, but of course there are vacuum fluctuations, from $\langle a^2 \rangle \neq 0$. When one considers quantum noise limits in interferometry, an open port (a term meaning one with just vacuum 'noise' incident) is a source of noise, as the zero point fluctuations of the vacuum field leak through the beamsplitter.

How do states transform via a beamsplitter? Well we can form states by letting a function of a and $a^\dagger$ acting on the vacuum. Let us consider a particular input state to the beam splitter,

$$|\psi\rangle_{\text{in}} = |0_1 1_2\rangle = a_2^\dagger|0_1 0_2\rangle \tag{5.174}$$

which refers to a vacuum state for mode 1 and a single photon state for 2. We take the beamsplitter to have $r = t = 1/\sqrt{2}$. Now, our output state can be determined by considering

$$a^\dagger = \frac{1}{\sqrt{2}}(c^\dagger - d^\dagger) \tag{5.175}$$

so

$$\begin{aligned}|\psi\rangle_{\text{out}} &= \frac{1}{\sqrt{2}}(c^\dagger - d^\dagger)|0_3 0_4\rangle \\ &= \frac{1}{\sqrt{2}}(|1_3 0_4\rangle - |0_3 1_4\rangle)\end{aligned} \tag{5.176}$$

This is a very interesting state, where we definitely have one photon in one arm and none in the other. This is an example of what we will call an entangled state in later chapters, meaning that the state cannot be written as a tensor product of states for the individual output modes. For these types of states, a measurement of system one yields some information about the other system. This state is what we call maximally entangled in that when we find a photon in mode 3 we *know* that there will be no photon in mode 4 and vice versa; we get total information about system 2 if we measure system 1.

What if we try two photons in one arm and none in the other? Let us see

$$|\psi\rangle_{\rm in} = |0_1 2_2\rangle = \frac{1}{\sqrt{2}} a_2^{\dagger 2} |0_1 0_2\rangle \tag{5.177}$$

This produces an output of the form

$$\begin{aligned} |\psi\rangle_{\rm out} &= \frac{1}{2}(c^{\dagger 2} + d^{\dagger 2} - 2c^\dagger d^\dagger)|0_3 0_4\rangle \\ &= \frac{1}{2}\left(\sqrt{2}\,|2_3 0_4\rangle - |1_3 1_4\rangle + \sqrt{2}\,|0_3 2_4\rangle\right) \end{aligned} \tag{5.178}$$

So, sometimes both photons go into mode 3, sometimes, both go into mode 4, and with half the probability, one goes into each mode. Here, the state generated is entangled, but is not maximally entangled. There is something special about the one photon state that is useful for making entanglement.

What if we put one photon in each input arm?

$$|\psi\rangle_{\rm in} = |1_1 1_2\rangle = a^\dagger b^\dagger |0_1 0_2\rangle \tag{5.179}$$

The output state is then

$$\begin{aligned} |\psi_{\rm out}\rangle &= \frac{1}{2}(c^{\dagger 2} + d^{\dagger 2} + c^\dagger d^\dagger - c^\dagger d^\dagger)|0_3 0_4\rangle \\ &= \frac{1}{\sqrt{2}}(|2_3 0_4\rangle + |0_3 2_4\rangle) \end{aligned} \tag{5.180}$$

Here, we again find a maximally entangled state, in that we know there are two photons in one output mode and none in the other. However, one can easily show that an input state of $|n_1 n_2\rangle$ does not give a simple state of the form $|(2n)_3 0_4\rangle + |0_3, (2n)_4\rangle$ but a distribution of all possible states of the form $|2n - m, m\rangle$. So, there is indeed something special about single photon states. There are two paths to generate one photon in either arm, that are indistinguishable. The amplitudes have opposite sign and destructively interfere. This is shown schematically in figure 5.24.

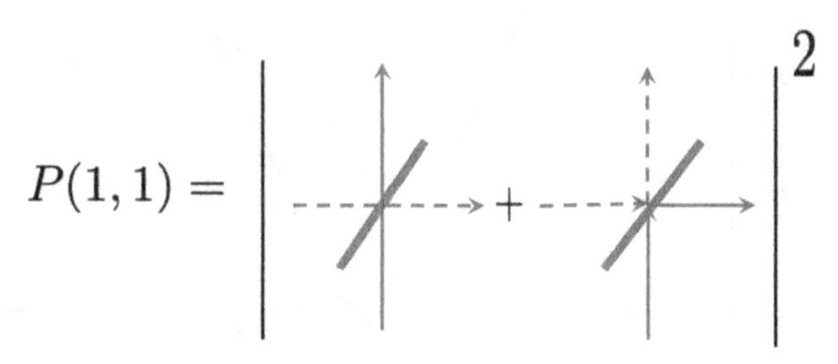

Figure 5.24. The two paths that lead to one photon in each output of the beam splitter, for a single photon incident in each input.

If we take the (spatially separate) outputs of a downconverter and direct them to the two inputs of a beam splitter, one never sees a coincidence count if the two path lengths are equal. One can unbalance the arms, and plotting the coincident counts versus path difference, one sees a dip when the arms are balanced, that is, no phase difference, as shown in figure 5.25. This is known as the Hong–Ou–Mandel effect, or HOM dip. If we do not have perfect single photon states at each input, we may still see a dip. A classical stochastic field can produce a dip of 50%. This can occur for two classical inputs whose frequencies are guaranteed to sum to the same value, via energy conservation, but the pair is not emitted at a random time, known as emission time freedom. A HOM dip of greater than 50% is a nonclassical effect. One can construct an interferometer, as shown in figure 5.26. When the two interferometer paths are balanced, one sees a rise in coincident counts.

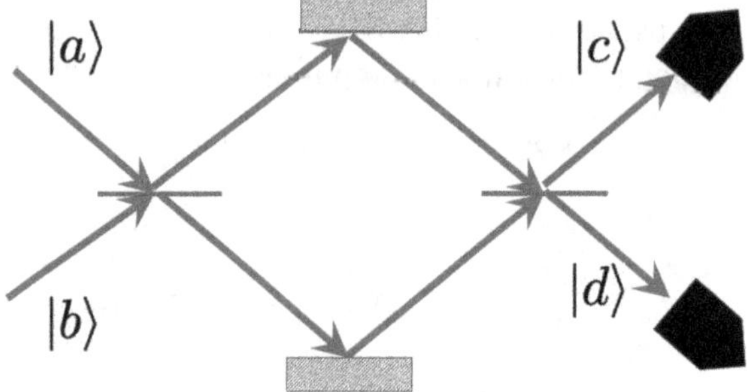

Figure 5.25. Diagram of a Hong–Ou–Mandel interferometer

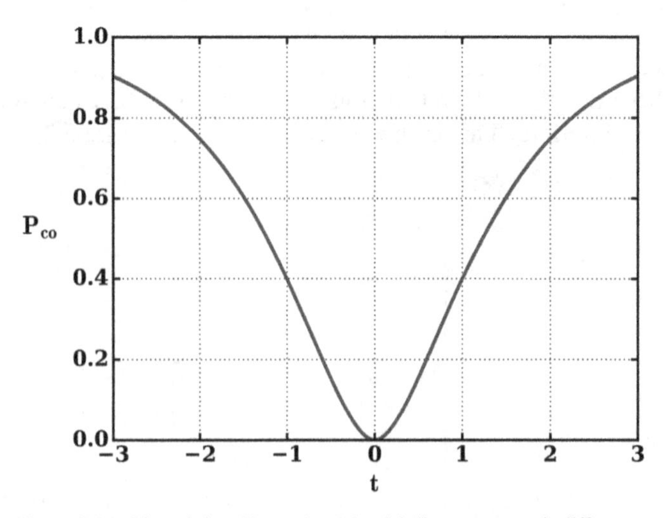

Figure 5.26. Plot of thee Hong–Ou–Mandel dip at zero path difference.

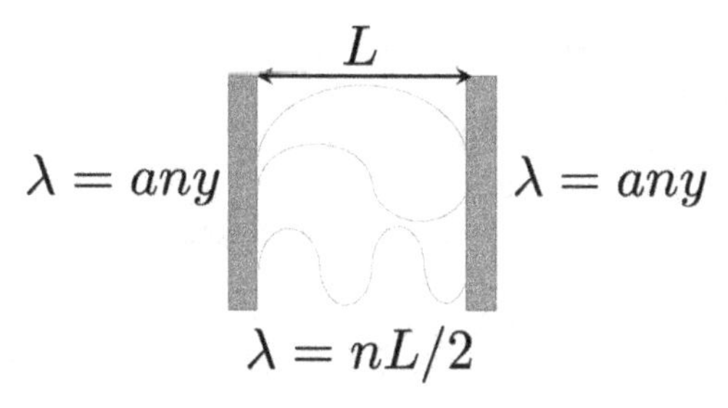

Figure 5.27. Unbalanced vacuum fluctuations in the Casimir effect.

What if we have vacuum into mode 1 and a coherent state into mode 2? This would mean an input state of

$$|\psi\rangle_{\text{in}} = |0_1, \alpha_2\rangle = D_2(\alpha)|0_1, 0_2\rangle \tag{5.181}$$

and it can be shown that

$$|\psi\rangle_{\text{out}} = |\alpha/\sqrt{2}, -\alpha/\sqrt{2}\rangle \tag{5.182}$$

which means that the amplitude is equally split between the two output modes, just as if we had a classical field in one input mode and nothing in the other.

5.10 Casimir effect

Another aspect is the Casimir effect. This is the fact that two conducting plates, effectively forming an optical cavity, will attract one another! The physical reason is that inside the cavity, only certain discrete modes exist, each with their corresponding vacuum fluctuations. Outside the cavity, there are fluctuations of all frequencies, as shown in figure 5.27. Each frequency/mode exerts a pressure due to its nonzero energy content, and the pressure is greater on the outsides of the cavity than on the inside, hence there is an attraction. We refer the interested reader to the nice treatment of Milonni [6]. Experiments have confirmed this effect's existence, albeit with a lot of theoretical work to go with heroic experiments. Seems that real metal plates are not perfect conductors and have all kinds of imperfections, but we seem to have a nice understanding of the basic effect as well as detailed calculations for particular experiments.

References

[1] Barnett S and Vaccaro J 2007 *The Quantum Phase Operator: A Review* (London: Taylor and Francis)

[2] Cohen-Tannoudji C, Grynberg G and Dupont-Roc J 1992 *Atoms–Photon Interactions: Basic Processes and Applications* (New York: Wiley)

[3] Cohen-Tannoudji C, Grynberg G and Dupont-Roc J 1997 *Photons and Atoms: Introduction to Quantum Electrodynamics* (New York: Wiley)

[4] Dirac P A M 1982 *The Principles of Quantum Mechanics* (Oxford: Clarendon)

[5] Loudon R 2000 *The Quantum Theory of Light* 3rd edn (Oxford: Oxford Science Publications)

[6] Milonni P W 1994 *The Quantum Vacuum: An Introduction to Quantum Electrodynamics* (New York: Academic)

[7] Noh J W, Fougeres A and Mandel L 1991 Measurement of the quantum phase by photon counting *Phys. Rev. Lett.* **67** 1426

[8] Pegg D T and Barnett S M 1989 Phase properties of squeezed states of light *Phys. Rev.* A **39** 1665

[9] Slusher R E, Hollberg L W, Yurke B, Mertz J C and Valley J F 1985 Observation of squeezed states generated by four-wave mixing in an optical cavity *Phys. Rev. Lett.* **55** 2409

[10] Susskind L and Glogower J 1964 Quantum mechanical phase and time operator *Phys. Phys. Fiz.* **1** 49

IOP Publishing

An Introduction to Quantum Optics
An open systems approach
Perry RIce

Chapter 6

Two-level atom coupled to a quantized field

6.1 Atom–field interaction in quantum optics

The atom–field interaction we consider is an electric dipole interaction with the same form we considered in the semi-classical theory of chapter 3 with an interaction energy given by

$$H_{\text{int}} = -\vec{d} \cdot \vec{E} \tag{6.1}$$

where $\vec{d}$ is the dipole moment of the atom and $\vec{E}$ is the electric field. Recall the details of the dipole moment of the atom depends on the atomic wavefunction

$$\vec{d} = \begin{pmatrix} \langle e|e\vec{r}|e\rangle & \langle e|e\vec{r}|g\rangle \\ \langle g|e\vec{r}|e\rangle & \langle g|e\vec{r}|g\rangle \end{pmatrix} \tag{6.2}$$

and the electric field is an operator

$$\vec{E} = \sqrt{\frac{\hbar\omega}{\epsilon_0 V}}(a + a^\dagger)\sin kz = \mathcal{E}_0(a + a^\dagger) \tag{6.3}$$

where V is the volume of the mode and k is the wavenumber of the mode. Assuming that the atom has no permanent dipole moment, the diagonal elements of the matrix dipole moment operator vanish and we can write the operator in terms of Pauli matrices

$$\begin{pmatrix} 0 & \mu_{eg} \\ \mu_{eg} & 0 \end{pmatrix} = \mu_{eg}(\sigma_+ + \sigma_-) \tag{6.4}$$

where μ_{eg} is the (real) dipole matrix element between the ground and excited states of the atom and σ_+ and σ_- are the same as in chapter 3. The transition matrix element is real for transitions that satisfy $\Delta m = 0$.

To get the operator form of the atom–field interaction, we use the dipole moment operator and electric field operator above. This gives

$$H_{\text{int}} = \hbar g_0 \sin kz(a + a^{\dagger})(\sigma_+ + \sigma_-) \tag{6.5}$$

$$= \hbar g_0 \sin kz(a\sigma_+ + a^{\dagger}\sigma_- + a\sigma_- + a^{\dagger}\sigma_+). \tag{6.6}$$

with the atom field coupling $g_0 \equiv \sqrt{(\mu_{eg}^2 \omega / 2\epsilon_0 V)}$. Considering each term of the interaction we see that $a\sigma_+$ destroys a photon and excites the atom, $a^{\dagger}\sigma_-$ creates a photon and lowers the atom, $a\sigma_-$ destroys a photon and lowers the atom, and $a^{\dagger}\sigma_+$ creates a photon and excites the atom. The first two terms correspond to absorption and emission and conserve energy while the last two terms do not conserve energy and describe virtual processes. Although these virtual processes give rise to small energy level shifts, they are not important in the present work and thus their contributions will be neglected. They correspond to the rapidly oscillating terms we neglected in the rotating wave approximation in our work that treated the field classically in chapter 4. This model was first considered by Jaynes and Cummings and is therefore known as the Jaynes–Cummings model. The final form of the interaction Hamiltonian is then

$$H_{\text{int}} = \hbar g_0 \sin kz(a\sigma_+ + a^{\dagger}\sigma_-) \tag{6.7}$$

or on changing a phase and mode function,

$$H_{\text{int}} = \imath\hbar g_0 \cos kz(a\sigma_+ - a^{\dagger}\sigma_-) \tag{6.8}$$

There is another way to see how the energy nonconserving terms can be neglected, let us consider the interaction picture. Here, we decompose the Hamiltonian into a 'bare' part that is independent of time plus a time dependent interaction term,

$$H = H_0 + H_I(t) \tag{6.9}$$

Recall from chapter 2 that the wave function evolves in time via the interaction Hamiltonian,

$$\dot{\psi} = -\frac{i}{\hbar} H_I(t)\psi \tag{6.10}$$

while operators also carry a time dependence

$$O(t) = e^{iH_0 t/\hbar} O(0) e^{-iH_0 t/\hbar} \tag{6.11}$$

For the annihilation and creation operators for the field, we have

$$a(t) = e^{i\omega a^{\dagger}at} a(0) e^{-i\omega a^{\dagger}at} = a(0)e^{-i\omega t} \tag{6.12}$$

$$a^{\dagger}(t) = e^{i\omega a^{\dagger}at} a^{\dagger}(0) e^{-i\omega a^{\dagger}at} = a^{\dagger}(0)e^{i\omega t} \tag{6.13}$$

To derive these results, we utilize the Baker–Hausdorff theorem. Consider $e^{aA}Be^{-aA}$ with A and B operators and a a scalar, and Taylor expand

$$
\begin{aligned}
e^{aA}Be^{-aA} &= (1 + aA + a^2A^2/2 + \cdots)B(1 - a^A + a^2A^2/2 + \cdots) \\
&= B + a(AB - BA) + (a^2/2)(A^2B - BA^2) + \cdots \\
&= B + a[A, B] + (a^2/2)(A^2B - ABA + ABA - BA^2) + \cdots \\
&= B + a[A, B] + (a^2/2)[A, [A, B]] + \cdots
\end{aligned}
\tag{6.14}
$$

It is also sometimes helpful to consider the commutator of a function $f(a^\dagger)$ with a. To begin, consider

$$
\begin{aligned}
[a, a^{\dagger 2}] &\equiv aa^{\dagger 2} - a^{\dagger 2}a \\
&= aa^{\dagger 2} - a^\dagger aa^\dagger + a^\dagger aa^\dagger - a^{\dagger 2}a \\
&= [a, a^\dagger]a^\dagger + a^\dagger[a, a^\dagger] \\
&= 2a^\dagger
\end{aligned}
\tag{6.15}
$$

Continuing in this manner, one can show

$$[a, a^{\dagger 3}] = 3a^{\dagger 2} \tag{6.16}$$

$$[a, a^{\dagger n}] = na^{\dagger n-1} \tag{6.17}$$

$$[a, f(a, a^\dagger)] = \frac{df}{da^\dagger} \tag{6.18}$$

and one can work out similar relations for commutators of the form $[a^\dagger, f(a)]$. In a similar manner, for the Pauli operators, one can show that

$$\sigma_+(t) = e^{i\omega\sigma_z t/2}\sigma_+(0)e^{-i\omega\sigma_z t/2} = \sigma_+(0)e^{i\omega t} \tag{6.19}$$

$$\sigma_-(t) = e^{i\omega\sigma_z t/2}\sigma_-(0)e^{-i\omega\sigma_z t/2} = \sigma_-(0)e^{-i\omega t} \tag{6.20}$$

Using these relations, the interaction Hamiltonian, in the interaction picture is

$$
\begin{aligned}
H_I(t) &= e^{i\omega(a^\dagger a+\sigma_z/2)t}H_I^{-i\omega(a^\dagger a+\sigma_z/2)t} \\
&= \hbar g_0 \cos(kz)(a\sigma_+e^{-i\Delta t} + a\sigma_-e^{i(\omega+\omega_0)t} \\
&\quad + a^\dagger\sigma_+e^{i(\omega+\omega_0)t} + a^\dagger\sigma_-e^{-i\Delta t})
\end{aligned}
\tag{6.21}
$$

whereas earlier we have $\Delta = \omega - \omega_0$. We see that dropping the rapidly oscillating terms is the same as dropping the energy nonconserving terms. We shall see in what follows that these terms do not yield any contribution to transitions.

What about the form of the wave function? We have seen that the quantized electromagnetic field is essentially a harmonic oscillator whose state can be expanded as a superposition of Fock states, and the atom has a two-state system with excited and ground states, so we can write

$$|\Psi\rangle = \sum_n \left(C_{g,n}(t)e^{-iE_{g,n}t/\hbar}|g, n\rangle + C_{e,n}(t)e^{-iE_{e,n}t/\hbar}|e, n\rangle\right) \tag{6.22}$$

As we have taken the zero of the atoms energy to lie between the two states, we have $E_{g,n} = \hbar\omega_0(n - 1/2)$ and $E_{e,n} = \hbar\omega_0(n + 1/2)$ where for now we have assumed that the atom and field are both resonant at ω_0. We have separated out the time dependence that occurs with no interaction (the exponential dependence) and that due to the interaction (the time dependence of the probability amplitudes). In this case, the Schrödinger equation

$$|\dot{\psi}\rangle = -\frac{i}{\hbar}H|\psi\rangle \tag{6.23}$$

gives us, for an atom at $z = 0$,

$$\dot{C}_{e,n} = -g\sqrt{n+1}\,C_{g,n+1} \tag{6.24}$$

$$\dot{C}_{g,n+1} = g\sqrt{n+1}\,C_{e,n} \tag{6.25}$$

Taking the time derivative of equation (6.24) and substituting it into equation (6.25), we find

$$\ddot{C}_{e,n} = -(n+1)g^2C_{e,n} \tag{6.26}$$

with solutions

$$C_{e,n}(t) = C_{e,n}(0)\cos(g\sqrt{n+1}\,t) - C_{g,n+1}(0)\sin(g\sqrt{n+1}\,t) \tag{6.27}$$

$$C_{g,n+1}(t) = C_{g,n+1}(0)\cos(g\sqrt{n+1}\,t) + C_{e,n}(0)\sin(g\sqrt{n+1}\,t) \tag{6.28}$$

One obtains exactly the same equations if one uses the wave function

$$|\psi\rangle_I = \sum_n \left(C_{g,n}|g,\, n\rangle + C_{e,n}|e,\, n\rangle\right) \tag{6.29}$$

and the interaction Hamiltonian equation (6.21).

For an atom initially in the ground state, $C_{g,n}(0) = 1$, we have

$$C_{e,n-1} = -\sin(g\sqrt{n}\,t) \tag{6.30}$$

and we see that at a later time, the probability of the atom being in the excited state while the field loses a photon is not zero; we have absorption. Notice that if $n = 0$ there is no absorption. What if the atom is initially in the excited state, $C_{e,n}(0) = 1$, we have

$$C_{g,n+1} = \sin(g\sqrt{n+1}\,t) \tag{6.31}$$

A seemingly extraordinary thing occurs; if the atom is in the excited state at $t = 0$ with no photons around ($n = 0$), there is still a probability for the atom to be in the ground state at a later time. By treating the field as a quantum operator, we find that spontaneous emission is contained automatically in the model.

Recall our semiclassical treatment of chapter 3,

$$\dot{C}_g = \frac{-i\Omega}{2} C_e \tag{6.32}$$

$$\dot{C}_e = \frac{-i\Omega}{2} C_g \tag{6.33}$$

with $\Omega = \mu_{eg} E_0 / \hbar$. One can make a connection via $\Omega/2 \rightarrow \Omega_{n+1}/2 \equiv g\sqrt{(n+1)} \approx \Omega_n/2$, which is consistent with $\hbar g/\mu_{eg}$ having units of electric field. We have ignored the difference between n and $n+1$, the semiclassical treatment is generally valid for large photon numbers, with the important distinction of needing to add spontaneous emission. It seems that we use $\Omega \rightarrow 2g\sqrt{n}$ for absorption, and $\Omega \rightarrow 2g\sqrt{n+1}$ for emission. The ratio of stimulated emission to spontaneous emission is n:1, and in a rate equation limit, one has a total emission rate of $n+1$ and a total absorption rate of n.

With a detuning of the driving laser from atomic resonance, $\Delta = \omega_L - \omega_0$, we have

$$\dot{C}_{e,n} = -g\sqrt{n+1}\, C_{g,n+1} - \imath \Delta C_{e,n} \tag{6.34}$$

$$\dot{C}_{g,n+1} = g\sqrt{n+1}\, C_{e,n} - \imath \Delta C_{g,n} \tag{6.35}$$

which leads to oscillatory solutions with frequencies $\Omega'_{n+1} = \sqrt{g^2(n+1) + \Delta^2}$, a generalized Rabi frequency.

For short times, we can approximate our results above, or use first-order perturbation theory, for example, if we start in the excited state of the atom with n photons, we find

$$C_{g,n+1} = g\sqrt{n+1}\, e^{-i\Delta t/2} \frac{\sin(\Delta t/2)}{\Delta t/2}\, t \tag{6.36}$$

which on resonance gives us

$$C_{g,n+1} = g\sqrt{n+1}\, t \tag{6.37}$$

where again we see that spontaneous emission occurs with no photons in the field. If instead we start in the ground state of the atom with n photons in the field we find

$$C_{e,n-1} = -g\sqrt{n}\, e^{-i\Delta t/2} \frac{\sin(\Delta t/2)}{\Delta t/2}\, t \tag{6.38}$$

where again we find no spontaneous absorption.

6.2 Wigner–Weisskopf approximation

Let us now consider the interaction of a two-level atom with many quantized field modes, as would be the case for an atom interacting with the vacuum modes around it [2]. In the interaction picture we have the Hamiltonian

$$H_{\text{int}} = i\hbar \sum_k \left(g_k \sigma_+ a_k e^{i\Delta_k t} - g_k^* a^\dagger \sigma_- e^{-i\Delta_k t} \right) \tag{6.39}$$

where k denotes all field modes that atom is coupled to. With the atom in the excited state initially, and all the field modes unoccupied, the only other states accessible via the above Hamiltonian are the atom in the ground state, and *one* of the field modes occupied with one photon. We then take the state vector

$$|\psi(t)\rangle = C_{e,0}(t)|e,\, 0\rangle + \sum_k C_{g,1k}|g,\, 1k\rangle \tag{6.40}$$

with initial conditions $C_{e,0} = 1$ and $C_{g,1k} = 0$. The Schrödinger equation (in the interaction picture) yields

$$\dot{C}_{e,0} = -\sum_k g_k^* e^{-i\Delta_k t} C_{g,1k} \tag{6.41}$$

$$\dot{C}_{g,1k} = g_k e^{i\Delta_k t} C_{e,0} \tag{6.42}$$

We can formally solve the second of these equations,

$$C_{g,1k} = g_k \int_0^t dt' e^{i\Delta_k t'} C_{e,0}(t') \tag{6.43}$$

We then substitute this into the equation for $C_{e,1}$ and find

$$\dot{C}_{e,0} = -\sum_k |g_k|^2 \int_0^t e^{-i\Delta_k (t-t')} C_{e,0}(t') \tag{6.44}$$

So far, we have not made any approximations beyond our usual ones (dipole, rotating wave). We now must address the mode distribution of the field. As we are interested in the case of free space, we realize that the spacing between modes $c/2L$ goes to zero and we have a continuum. The discrete sum over modes can be turned into an integral and switch to polar coordinates so that $d^3\vec{k} = 2dk_x dk_y dk_z = k^2 \sin(\theta) d\theta d\phi dk$ where the factor of 2 is due to the two possible polarizations. We have $V = L^3$ as the quantization volume. This leaves us with

$$\dot{C}_{e,0} = -\frac{2V|g_k|^2 \omega}{(2\pi)^3} \int_0^{2\pi} d\phi \int_0^{\pi} d\theta \sin(\theta) \cos^2(\theta) \int_0^{\infty} dk k^2 |g_k|^2 \int_0^t e^{i\Delta(t-t')} C_{g,1}(t') \tag{6.45}$$

The factor of $(L/2\pi)^3 = V/(2\pi)^3$ stems from the volume of the unit cell in phase space. The ϕ integral merely yields a factor of 2π, and the θ integral is an average of $\cos^2(\theta)$ over three dimensions, which yields 1/3. This leaves us with

$$\dot{C}_{e,0} = -\frac{4|\mu_{eg}|^2}{6(2\pi)^2\epsilon_0\hbar c^3}\int_0^\infty d\omega\omega^3\int_0^t e^{i\Delta(t-t')}C_{e,0}(t') \tag{6.46}$$

where we have used the relations $\omega = ck$ and $g^2 = \omega|\mu_{eg}|^2/2\hbar\epsilon_0 V$. Now, we make an approximation, taking the function ω^3 to be smoothly varying over the range of frequencies in the integral (basically the frequencies involved in the time development of $C_{e,0}(t)$ and the complex exponential. We replace it with ω_0^3 which then leaves us with

$$\dot{C}_{e,0} = -\frac{4|\mu_{eg}|^2\omega_0^3}{6(2\pi)^2\epsilon_0\hbar c^3}\int_0^\infty d\omega\int_0^t e^{-i\Delta(t-t')}C_{e,0}(t') \tag{6.47}$$

The factor of $\cos^2(\theta)$ stems from the dot product between the field mode and the induced dipole moment. At this point, we assume that $C_e(t)$ is slowly varying compared to the oscillating exponential and pull it outside the integral. The frequency integral is then

$$\int_{-\infty}^\infty d\omega e^{i\Delta(t-t')} = 2\pi\delta(t-t') \tag{6.48}$$

where we have extended the lower limit of the integral to $-\infty$. The δ function kills the time integral, leaving us with

$$\dot{C}_{e,0} = -\frac{\gamma}{2}C_{e,0} \tag{6.49}$$

with

$$\gamma = \frac{\omega_0^3|\mu_{eg}|^2}{6\pi\epsilon_0\hbar c^3} \tag{6.50}$$

which leads to exponential decay

$$C_{e,0}(t) = e^{-\gamma t/2} \tag{6.51}$$

for the probability amplitude, or for the probability we have

$$P_{e,0} = e^{-\gamma t} \tag{6.52}$$

This result is the same that we obtained when we considered spontaneous emission via rate equations in chapter 3, where we found the Einstein A coefficient,

$$A = \gamma = \frac{\omega_0^3|\mu_{eg}|^2}{6\pi\epsilon_0\hbar c^3} \tag{6.53}$$

This treatment has a theme that we will see again when we consider a more general method of treating spontaneous emission and other dissipative processes in chapter 7. The density of states is proportional to an integral of ω^2 over ω, and as such it tends to blow up, i.e. diverge. This simply stems from the fact that there are an infinite amount of modes. The coupling of the atom to each mode is small, however, as $g^2 \propto 1/V \to 0$, and also many of the modes have some amount of detuning. So, we sort of have a large number (number of modes) multiplying a small number (the coupling to each individual mode) that results in a finite result. In our formal work in chapter 7, we will see that this is a type of Born approximation. Also, we are making assumptions about the fact that the atom changes slowly over the time scale that the environment (the vacuum modes) can respond that the environment reacts very fast, and this is known as the Markov approximation. This is a very good approximation for systems that oscillate at optical frequencies, and the typical theory of open (i.e. dissipative) systems makes use of the combined Born–Markov approximations.

6.3 Cavity modified spontaneous emission

If we modify the density of states from that of the free space distribution, we can alter the spontaneous emission rate from its free space value. In a laser system, we might wish to enhance spontaneous emission into the cavity mode (which generates stimulated emission) and get rid of spontaneous emission out the sides of the cavity (which is an energy loss mechanism). The simplest change in the density of states is to make it zero! This could result from a cavity where all linear dimensions are less than $\lambda_0/2$, and so there are no modes for the atom to radiate into. This would result in a complete suppression of spontaneous emission.

6.4 Dressed states reprise

One choice of basis for dealing with atom–field systems is just the tensor product of the Fock states with the two atomic states. These are known as the bare states and are shown in figure 6.1.

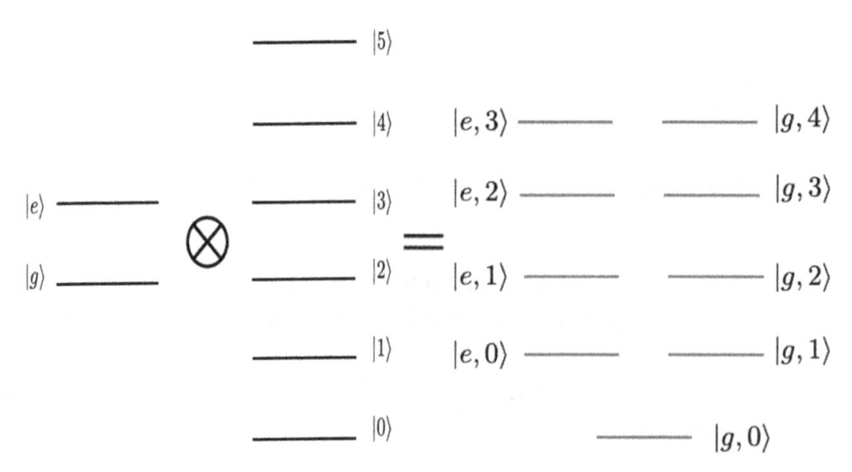

Figure 6.1. Bare states of a field–atom system.

Recall that for an atom interacting with a classical field, we considered the Hamiltonian

$$H = \frac{\hbar\Omega}{2}\sigma_z + i\hbar\Omega(\sigma_+ - \sigma_-) \tag{6.54}$$

with $\Omega = |\mu|_{eg}E_0/\hbar$ and found

$$|+\rangle = \cos(\Theta/2)|e\rangle + \sin(\Theta/2)|g\rangle \tag{6.55}$$

$$|-\rangle = -\sin(\Theta/2)|e\rangle + \cos(\Theta/2)|g\rangle \tag{6.56}$$

with

$$\tan(\Theta) = \frac{\Omega}{\Delta} \tag{6.57}$$

with energies

$$E_\pm = \frac{\hbar}{2}(-\Delta \pm \Omega') \tag{6.58}$$

Can we also consider the eigenstates of the full atom–field Hamiltonian with a quantized field? Sure! The Hamiltonian is, ignoring the constant zero point energy of the field,

$$H = \hbar\omega a^\dagger a + \frac{\hbar\omega_0}{2}\sigma_z + i\hbar g(a\sigma_+ - a^\dagger\sigma_-) \tag{6.59}$$

The matrix representation in the basis defined by $|g, n\rangle$ and $|e, n\rangle$ is block diagonal. The ground state energy is still $\hbar\omega/2$ even with the interaction and so that is the upper left entry. The rest of the Hamiltonian is 2×2 blocks, as the interaction Hamiltonian only couples states $|g, n + 1\rangle$ and $|e, n\rangle$ for $n \geqslant 1$. These blocks are

$$H = \hbar\begin{pmatrix} n\omega + \omega_0/2 & g\sqrt{n+1} \\ -g\sqrt{n+1} & (n+1)\omega - \omega_0/2 \end{pmatrix} \tag{6.60}$$

The eigenvalues (allowed energies) are

$$E_\pm = (n + 1/2)\hbar\omega \pm \frac{1}{2}\hbar\sqrt{\Delta^2 + 4g^2(n+1)} \tag{6.61}$$

This is equivalent to the semiclassical theory if we make the replacement $2g\sqrt{n+1} \rightarrow \Omega$ and ignore the energy of the field mode $n\hbar\omega$ as before. The eigenstates are

$$|n, +\rangle = \cos(\Theta_n/2)|e, n\rangle + \sin(\Theta_n/2)|g, n+1\rangle \tag{6.62}$$

$$|n, -\rangle = -\sin(\Theta_n/2)|e, n\rangle + \cos(\Theta_n/2)|g, n+1\rangle \tag{6.63}$$

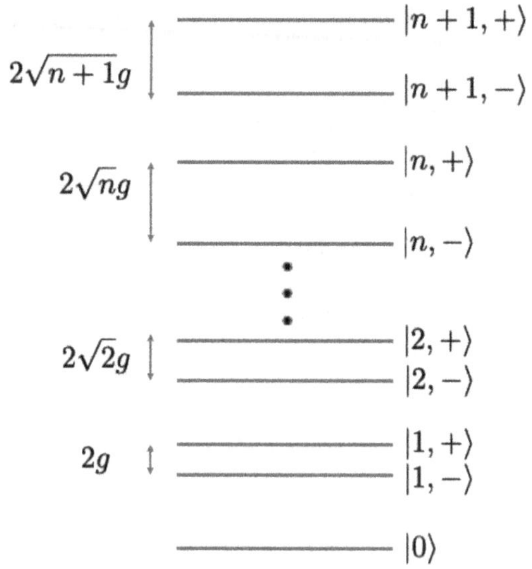

Figure 6.2. Dressed state ladder of Jaynes–Cummings states.

with

$$\tan(\Theta_n) = \frac{\Omega_n}{\Delta} = \frac{2g\sqrt{n+1}}{\Delta} \tag{6.64}$$

which once again can be turned into the semiclassical results with $\Omega_n \approx \Omega_{n+1} \to \Omega$. Again, on resonance we have

$$|n, +\rangle = \sqrt{\frac{1}{2}}(|e, n\rangle + |g, n+1\rangle) \tag{6.65}$$

$$|n, -\rangle = \sqrt{\frac{1}{2}}(|g, n+1\rangle - |e, n\rangle) \tag{6.66}$$

The energy levels, known as the Jaynes–Cummings ladder, are shown below in figure 6.2. Of particular note is that the two one-excitation states have a splitting of $\Delta E = 2\hbar g$. Recall that an atom in the excited state with no photons in the field would be represented by the state

$$|e, 1\rangle = \sqrt{\frac{1}{2}}(|1, +\rangle - |1, -\rangle) \tag{6.67}$$

and so it is in a superposition of the two one-excitation states. Spontaneous emission from this state would result in *two* spectral lines, which are known as the vacuum-Rabi doublet, or vacuum-Rabi splitting, $\Delta\omega = 2g$. The 'vacuum' part of this moniker is due to the fact that the initial state of the field has no photons in it, it is in the vacuum state. But note that if the atom were in the ground state initially with one photon in the field, the splitting would still occur, and so there is no magical mystical property of the vacuum involved. In fact, as we will show later, the phenomena of vacuum-Rabi splitting can be obtained from coupled oscillators. For these reasons we prefer the term 'normal-mode splitting'. Now, if one observes splittings higher up the Jaynes–Cummings ladder, say $\Delta\omega = g(2\sqrt{2} - 2)$ in addition to the $\Delta\omega = 2g$ splitting, then one has observed the discrete nature of the coupled atom–field system when it is quantized. This has been observed in the time domain by the group of Haroche. As the normal-mode splitting is essentially a classical phenomenon, it can be observed in solids at room-temperature.

6.5 Heisenberg equations of motion

Let us consider the quantized atom–field Hamiltonian

$$H = \hbar\omega a^\dagger a + \frac{\hbar\omega_0}{2}\sigma_z + i\hbar g(a^\dagger\sigma_- - a\sigma_+) \tag{6.68}$$

Let us now calculate Heisenberg equations for the operators of interest, using

$$\dot{O} = \frac{i}{\hbar}[H, O] \tag{6.69}$$

Let us start with the annihilation operator for the field,

$$\begin{aligned} \dot{a} &= i\omega[a^\dagger a, a] - g[a^\dagger, a]\sigma_- \\ &= -i\omega a + g\sigma_- \end{aligned} \tag{6.70}$$

similarly we find

$$\dot{a}^\dagger = i\omega a^\dagger + g\sigma_+ \tag{6.71}$$

$$\dot{\sigma}_- = i\omega_0\sigma_- + ga^\dagger\sigma_+ \tag{6.72}$$

$$\dot{\sigma}_z = -2g(a^\dagger\sigma_- - a\sigma_+) \tag{6.73}$$

Evidently, we must continue on to consider quantities like $d(a\sigma_+)/dt$. This then couples to other operator products, for example

$$\frac{d(a\sigma_+)}{dt} = -i(\omega + \omega_0)a\sigma_+ + 2ga\sigma_z \tag{6.74}$$

and then on and on we go. The system of equations for the operators does not close, so to solve things we would need some type of approximation or truncation scheme. Of some interest is the equation of motion for the photon number,

$$\frac{d(a^\dagger a)}{dt} = g(a^\dagger\sigma_- - a\sigma_+) \tag{6.75}$$

where we note that we then have

$$\frac{d(a^\dagger a + \sigma_z/2)}{dt} = \frac{dC_1}{dt} = 0 \tag{6.76}$$

which is a statement of energy conservation. There is another conserved operator,

$$C_2 = \frac{\Delta\sigma_z}{2} + ig(a^\dagger\sigma_- - a\sigma_+) \tag{6.77}$$

Now, we can relate these back to the Maxwell–Bloch equations of chapter 3, by considering expectation values of the operator equations above, along with a factorization of operator products

$$\langle\dot{a}^\dagger\rangle = i\omega\langle a^\dagger\rangle + g\langle\sigma_+\rangle - \kappa\langle a\rangle \tag{6.78}$$

$$\langle\dot{\sigma}_-\rangle = i\omega_0\langle\sigma_-\rangle + g\langle a^\dagger\rangle\langle\sigma_+\rangle - \frac{\Gamma}{2}\langle\sigma_-\rangle \tag{6.79}$$

$$\langle\dot{\sigma}_z\rangle = -2g(\langle a^\dagger\rangle\langle\sigma_-\rangle - \langle a\rangle\langle\sigma_+\rangle) - \gamma(\langle\sigma_z\rangle + 1) \tag{6.80}$$

where we have approximated $\langle a^\dagger\sigma_-\rangle \approx \langle a^\dagger\rangle\langle\sigma_-\rangle$ and added phenomenological decay terms that we will derive later. If we make the correspondence $g\langle a\rangle = g\langle a^\dagger\rangle \approx \Omega/2$, we basically have the Maxwell–Bloch equations we discussed earlier. Of course, this factorization results in the semiclassical model, but is indeed incorrect as one must use the fully quantum mechanical model.

6.6 Collapse and revivals of population inversion

Let us consider the population inversion of a two-level atom interacting with a quantized field [1]. This is given by

$$w(t) = \sum_n \left(|C_{e,n}(t)|^2 - |C_{g,n}(t)|^2\right) \tag{6.81}$$

where the summation is over the number states of the field. If we have a definite number of photons in the field, $n = m$, and the initial state of the atom is the excited state, then this simplifies to

$$\begin{aligned} w(t) &= \cos^2(g\sqrt{n+1}\,t) - \sin^2(g\sqrt{n+1}\,t) \\ &= \cos(2g\sqrt{n+1}\,t) \end{aligned} \tag{6.82}$$

where we see Rabi oscillations due to the exchange of energy between the field and the atom. If instead the state of the field is not a Fock state, but initially has some distribution over Fock states, independent of the atoms state. In that case the initial state vector is

$$|\psi(0)\rangle = \sum_n (C_n|n\rangle)\left(C_e|e\rangle + C_g|g\rangle\right) \tag{6.83}$$

we then have

$$w(t) = \sum_n P_n \cos(2g\sqrt{n+1}\,t) \tag{6.84}$$

where we have taken $P_n = |C_n|^2$. A plot of the inversion versus time is shown in figure 6.3, for an initial photon distribution for a coherent state. The early time evolution showing the collapse is presented in figure 6.4.

We notice that the inversion oscillates at a mean Rabi frequency $\Omega = g\sqrt{\bar{n}} = g|\alpha|$, but eventually decays, or 'collapses' to zero. At a later time these oscillations recur, or 'revive'. This is due entirely to the quantized nature of the field. The sum in equation (6.84) is a sum of cosines whose frequencies are not integral multiples of one another. Another way to say this is that equation (6.84) is *not* a Fourier series, that the population is not a periodic function. The different terms oscillate at different frequencies and after some time, they are out of phase and destructive

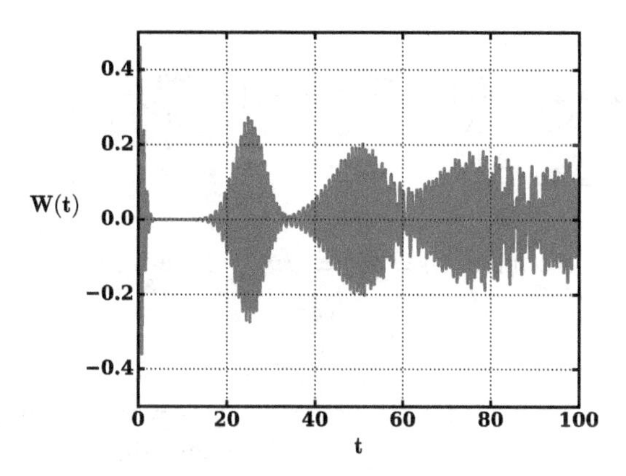

Figure 6.3. Collapse and revival of atomic inversion for initial coherent field state $\alpha = 4$.

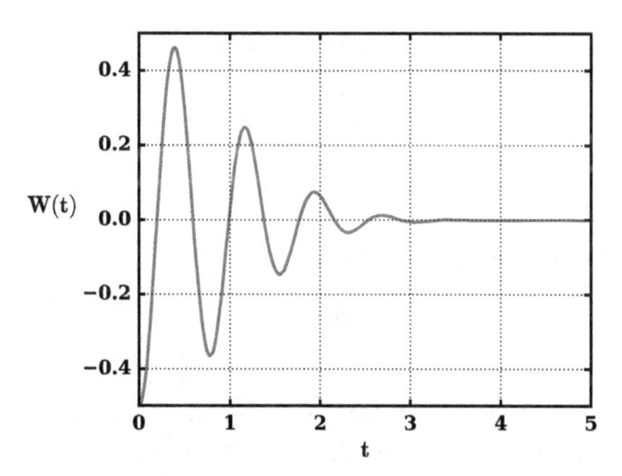

Figure 6.4. Collapse and revival of atomic inversion for initial coherent field state $\alpha = 4$. Only early rimes to show collapse.

interference occurs. This would occur whether we used a sum over discrete photon numbers (as quantum mechanics tells us we must) or an integral to sum over a continuum of photon numbers (as some type of classical theory with no quantization of field amplitudes would allow). However, at a later time, constructive interference occurs, and this revival is a signature of a quantized field, i.e. that only discrete values of the photon number are allowed in the sum. To see this, let us examine the time scale of the collapses and revivals.

We can get a rough approximation for the collapse time by considering two terms of the series, one that corresponds to $n = \bar{n} + \Delta n$ and $n = \bar{n} - \Delta n$. These give us two Rabi frequencies that bracket the range of frequencies in the sum equation (6.84). When the phase difference between the two contributions is π, we will start to see cancellations in the sum, and a 'collapse' occurs at time t_c given by

$$2gt_c(\sqrt{\bar{n} + \Delta n + 1} - \sqrt{\bar{n} - \Delta n + 1}) = \pi \tag{6.85}$$

Now, we assume that the average photon number is large and expand the square roots obtaining

$$\begin{aligned} \pi &= 2gt_c\sqrt{\bar{n}}\left(1 + \frac{\Delta n}{\bar{n}} + \frac{1}{\bar{n}} - 1 + \frac{\Delta n}{\bar{n}} - \frac{1}{\bar{n}}\right) \\ &= 4gt_c\frac{\Delta n}{\sqrt{\bar{n}}} \end{aligned} \tag{6.86}$$

For a coherent state, we have $\Delta n = |\alpha| = \sqrt{\bar{n}}$ and we have

$$t_c = \frac{\pi}{4g} \tag{6.87}$$

which is independent of $\bar{n} = |\alpha|^2$. The revivals occur when the frequencies involved are all commensurate again, meaning that the phase difference between adjoining terms is an integer times 2π, at the mth revival time t_r^m

$$\begin{aligned} 2\pi m &= 2gt_r^m(\sqrt{\bar{n} + 1} - \sqrt{\bar{n}}) \\ &= 2gt_r^m\sqrt{\bar{n}}(\sqrt{1 + 1/\bar{n}} - 1) \\ &\approx 2gt_r^m\sqrt{\bar{n}}\left(1 + \frac{1}{2\bar{n}} - 1\right) \\ &= gt_r^m/\sqrt{\bar{n}} \end{aligned} \tag{6.88}$$

which leads to a revival time of the order

$$t_r^m = 2\pi m\sqrt{\bar{n}}/g \tag{6.89}$$

which agrees with the results of the full summation in figure 6.3. We have expanded the square root which is appropriate for large $\bar{n}$. We can do a more rigorous derivation, let us write $n = \bar{n} + \delta n$, where δn is the difference of n from the average number and is not to be confused with the standard deviation of the photon distribution Δn. The main part of this is the expansion of the square root,

$$
\begin{aligned}
\sqrt{n+1} &= \sqrt{\bar{n}+\delta n+1} \\
&= \sqrt{\bar{n}}\sqrt{1+\frac{\delta n}{\bar{n}}+\frac{1}{\bar{n}}} \\
&= \sqrt{\bar{n}}\left(1+\frac{\delta n}{2\bar{n}}\right)
\end{aligned}
\tag{6.90}
$$

The inversion then becomes

$$
\begin{aligned}
w(t) &= e^{-\bar{n}}\sum_n \frac{\bar{n}^n}{n!}\cos\left(g\sqrt{\bar{n}}\left(1+\frac{\delta n}{2\bar{n}}\right)\right) \\
&= \frac{1}{2}e^{-\bar{n}}\sum_n (e^{2igt\sqrt{\bar{n}+1}}e^{igt(n-\bar{n})/\sqrt{\bar{n}}} + e^{-2igt\sqrt{\bar{n}+1}}e^{-igt(n-\bar{n})/\sqrt{\bar{n}}})
\end{aligned}
\tag{6.91}
$$

The sums can be done using

$$
\sum_n \frac{\bar{n}^n}{n!}e^{igt(n-\bar{n})/\sqrt{\bar{n}}} = \exp(\bar{n}e^{igt/\sqrt{\bar{n}+1}})
\tag{6.92}
$$

For short times, we find that

$$
e^{igt/\sqrt{\bar{n}+1}} \simeq 1 + \frac{igt}{\sqrt{\bar{n}+1}} - \frac{g^2t^2}{2(\bar{n}+1)}
\tag{6.93}
$$

and we then have for the inversion (for short times)

$$
w(t) - \cos\left(2gt\sqrt{\bar{n}+1}\right)\exp\left(-\frac{g^2t^2\bar{n}}{2(\bar{n}+1)}\right)
\tag{6.94}
$$

which has a Gaussian dependence with decay time

$$
t_c = \frac{1}{g}\sqrt{\frac{2(\bar{n}+1)}{\bar{n}}} \approx \frac{\sqrt{2}}{g}
\tag{6.95}
$$

which is consistent with our previous estimate. In figure 6.3, we show an example of a collapse and revivals for $\alpha = 4$. A blowup of early times to show the collapse is shown in figure 6.4.

In figure 6.5, we illustrate the quantum collapse and revival with a phasor diagram. All of the phasors have incommensurate frequencies, but they are discrete in the quantum case. As indicated above, a time will come when they do indeed arrive back in phase, leading to a revival.

By contrast in the classical case, figure 6.6, all frequencies occur, and after the collapse there is never a time when all of the phasors recur, hence no revival. The revival is a direct indication of the quantum nature of the electromagnetic field.

Discrete frequencies (Quantum)

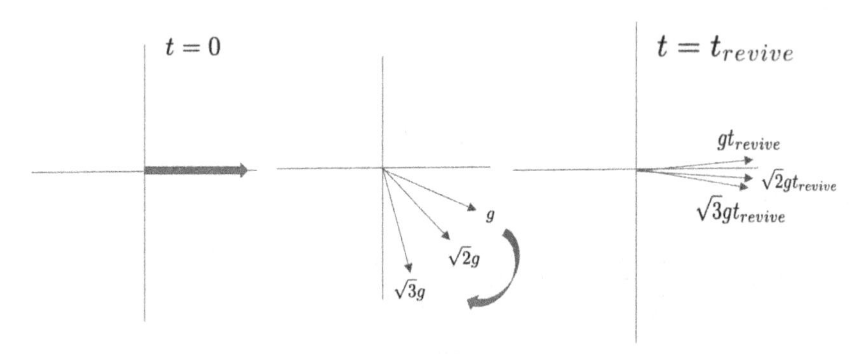

Figure 6.5. Phasor diagram of a few quantum states rotating at discrete incommensurate frequencies, leading to a collapse and revival.

Continuous frequencies (Classical)

Collapse, no revival

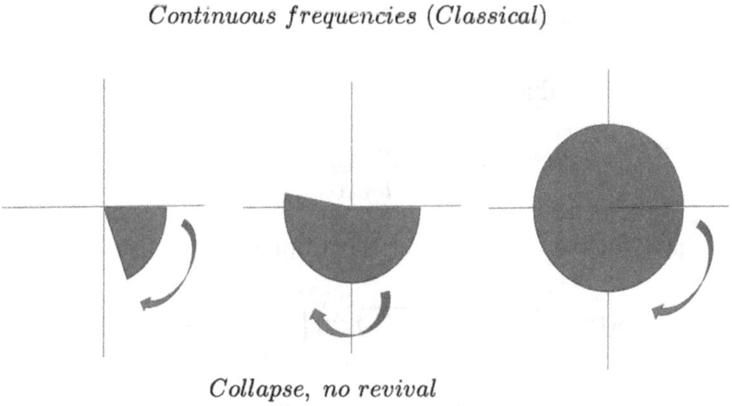

Figure 6.6. Phasor diagram of a few quantum states rotating at discrete incommensurate frequencies, leading to a collapse and revival.

6.7 Vacuum fluctuations and radiation reaction

Are vacuum fluctuations real? Yes indeed. An electron is surrounded by a cloud of virtual electron–positron pairs that pop into and out of existence. This is allowed by the time energy uncertainty principle, $\Delta E \Delta t \geqslant \hbar/2$. To create such a pair, one must have an energy fluctuation of at least $2mc^2$, so the longest the fluctuation can last is $\Delta t = \hbar/2mc^2$. Pairs with kinetic energy last for a shorter time. When one combines special relativity with quantum mechanics, one is immediately led to various new effects. First is the intrinsic spin of the electron, this fact falls directly out of the construction of the Dirac equation. Secondly, the Dirac equation also predicts antimatter, particles of opposite charge, but otherwise identical. They are created in pairs and annihilate to yield two counterpropagating photons. Finally, the necessity

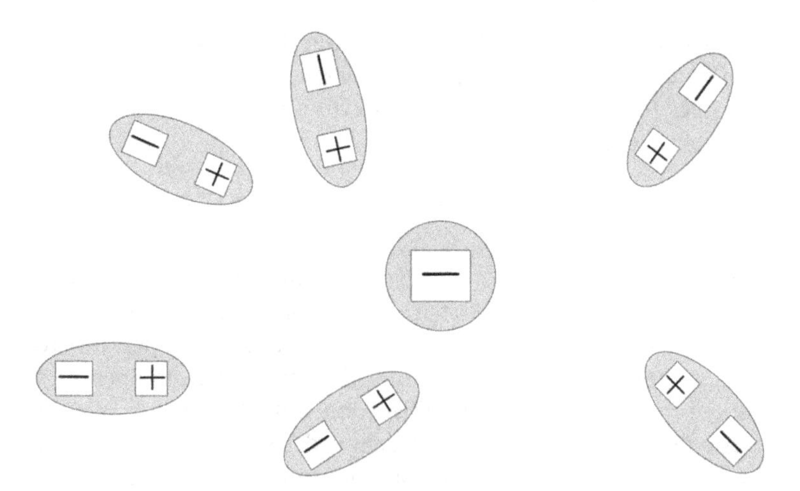

Figure 6.7. Cloud of virtual pairs around a bare particle.

of antimatter, and the time energy uncertainty principle, means we must have this virtual cloud of particle–antiparticle pairs, as shown in figure 6.7. The marriage of SR and QM is known as quantum field theory, and there are no single particle states. This virtual cloud acts as a dielectric medium that the electron finds itself in. The effects of this cloud show up as Lamb shifts, small changes in the resonant frequency of a transition. Naive calculations show that the mass and charge of this single electron (known as a bare electron) need to be infinite. However, the charge and mass we measure in the lab is of course not infinite. *But* in the lab we see the electron and the cloud, the real electron as it were. One can find a logically consistent procedure for doing calculations with the measured charge and mass. It means we should *start* with a theory for the electron plus cloud and not pretend that single particles exist. For very high energy probes, you do not measure the same values as at low energies, you probe deeper into the cloud essentially. These Lamb shifts are typically not mentioned in quantum optics. We cannot see the bare transition, so we take the experimental location of the spectral line to be the bare transition frequency plus the Lamb shift. One can predict the size of these shifts for simple atoms like hydrogen and get agreement to nine decimal places or more. So, it is an ugly procedure in some ways, but it is the best theory we have! It leads to atomic clocks that may only lose a fraction of a second over the lifetime of the Universe. There are also virtual pairs of $W^+ - W^-$ and pairs of Z^0 particles, the exchange bosons for the weak force. The electromagnetic and weak interactions are unified in a gauge theory using $U(1) \otimes SU(2)$. The particle we call the photon is actually a mixture of the $U(1)$ boson and the uncharged $SU(2)$ boson. Similarly, the Z^0 ia a mixture of the two fundamental bosons. If that were not the case, the theories would not be unified, but just stapled together. In fact, the most precise measurements of the mass of the vector bosons $W^\pm$, Z^0 come from precise measurements of Lamb shifts.

What about spontaneous emission? Why does an excited atom eventually decay? If it is due to the 'jostling' of vacuum fluctuations, why is there no spontaneous absorption? To give an answer to this, we consider the rolds of vacuum fluctuations, and radiation reaction in spontaneous processes.

Let us consider the interaction of a two-level atom with a collection of modes, which are in their vacuum state

$$H = \hbar\omega_0\sigma_z + \sum_k \hbar\omega_k a_k^\dagger a_k + \sum_k g_k(\sigma_+ a_k + a_k^\dagger \sigma_-) \tag{6.96}$$

where the index k runs over all wavevectors and polarizations. The resonant frequency of the atom is ω_0, and the frequency of the field modes is denoted by ω_k. We have written this in normal ordering, the annihilation operators to the right, and creation operators to the left. Now, since the field and atom operators commute at equal times, it makes absolutely no difference in the final result. We will preserve this order throughout the calculation, different orderings will give the same result for the Einstein A coefficient, but the (possible) interpretation will change.

The Heisenberg equations of motion are, preserving the order

$$\dot{\sigma}_- = -i\omega_0\sigma_- + \sum_k g_k a_k^\dagger \sigma_z \tag{6.97}$$

$$\dot{\sigma}_z = -2\sum_k g_k(\sigma_+ a - a_k^\dagger \sigma_-) \tag{6.98}$$

$$\dot{a}_k = -i\omega_k + g_k(\sigma_+ + \sigma_-) \tag{6.99}$$

We formally integrate the field operator equation to find

$$a_k(t) = a_k(0)e^{-i\omega_k t} + g_k \int_0^t dt' \sigma_- e^{-i\omega_k(t-t')} \tag{6.100}$$

and the lowering operator to find

$$\sigma_-(t) = -i\omega_0\sigma_- - \sum_k g_k \int_0^t dt' a_k e^{-i\omega_-(t-t')} \tag{6.101}$$

These solutions enter into the equation for the inversion operator multiplied by g_k. So, to keep the calculation at second order altogether in g_k, for purposes of substitution into that equation we must use

$$\sigma_-(t') = \sigma_-(t)e^{-i(\omega_0(t'-t))} \tag{6.102}$$

and

$$a(t') = a(t)e^{-i(\omega_k(t'-t))} \tag{6.103}$$

to find

$$a_k(t) = a_k(0)e^{-i\omega_k t} + g_k\sigma_-(t)\int_0^t dt' e^{-i(\omega_k - \omega_0)(t-t')} \tag{6.104}$$

$$=a_k(0)e^{-i\omega_k t} + \pi g_k \sigma_-(t) \tag{6.105}$$

and

$$\sigma_-(t) = \sigma_-(0)e^{-i\omega_0 t} - \sum_k g_k a_k(t)\int_0^t dt' e^{-i(\omega_k-\omega_0)(t-t')} \tag{6.106}$$

$$=-i\omega_0\sigma_- - \pi\sum_k g_k a_k(t) \tag{6.107}$$

This is the same approximation we have used in deriving the Wigner–Weisskopf result. Substituting these solutions into the equation for the inversion operator, we find

$$\dot{\sigma}_z = \left[-2\sum_k(\sigma_-(0)a_k(0)e^{-i(\omega_0-\omega_k)t}\right] \tag{6.108}$$

$$-2\pi^2\sum_k \pi^2 g_k^2(a_k^\dagger(t)a(t)) \tag{6.109}$$

$$\left.-2\pi^2\sum_k \pi^2 g_k^2(\sigma_+(t)\sigma_-(t))\right] \tag{6.110}$$

$$+\ \text{h.c.} \tag{6.111}$$

Taking an expectation value over the vacuum field state and using product rules for the Pauli operators, we find

$$\dot{\sigma}_z = 0 - 4\beta(\sigma_z + 1) \tag{6.112}$$

with

$$\beta = \pi^2\sum_k g_k^2 \tag{6.113}$$

Converting the sum to an integral over field couplings in the continuum limit, we find

$$\dot{\sigma}_z = 0 - \gamma(\sigma_z + 1) \tag{6.114}$$

where γ is the usual free space spontaneous emission rate, the Einstein A coefficient.

It would appear that the vacuum field makes no contribution as $\langle a_k^\dagger a_k\rangle = 0$, hence the spontaneous emission comes from the radiation reaction, $\gamma = \gamma_{RR}$

Let us consider using

$$\dot{\sigma}_z = -2\sum_k g_k(a\sigma_+ - a_k^\dagger\sigma_-)a^\dagger \tag{6.115}$$

As field and atom operators commute, we will of course obtain the same result. Going through the previous procedure, we find

$$\dot{\sigma}_z = \left[-2\sum_k (\sigma_-(0)a_k(0)e^{-i(\omega_0-\omega_k)t}) \right. \tag{6.116}$$

$$-2\pi^2 \sum_k \pi^2 g_k^2 (a_k(t)a_k^\dagger(t)) \tag{6.117}$$

$$\left. -2\pi^2 \sum_k \pi^2 g_k^2 (\sigma_-(t)\sigma_+(t)) \right] \tag{6.118}$$

$$+ \text{h.c.} \tag{6.119}$$

Taking expectation values with respect to the vacuum field yields

$$\dot{\sigma}_z = 0 - \gamma(\sigma_z - 1) - 2\gamma\sigma_z \tag{6.120}$$

where the first term comes from $\sigma_-\sigma_+ = (1/2)(\sigma_z - 1)$ and $\langle a_k^\dagger a_k \rangle = 1$.

The final result is of course

$$\dot{\sigma}_z = -\gamma(\sigma_z + 1) \tag{6.121}$$

but now we have $\gamma = 2\gamma_{VF} - \gamma_{RR}$, and so our interpretation can be different; spontaneous emission comes from an interplay of radiation reaction and vacuum fluctuations. As Cohen-Tannoudji pointed out, using a symmetrical ordering for field operators yields $\gamma = \gamma_{VF}$ and it seems that spontaneous emission is purely a vacuum fluctuation result. He argues that since symmetrically ordered operators are Hermitian, one could in principle measure them separately, and that the vacuum fluctuations are best considered as the source for spontaneous emission. What we do know for sure is that we obtain the Einstein *A* coefficient no matter what order we use for commuting atom and field operators!

References

[1] Eberly J H, Narozhny N B and Sanchez-Mondragon J J 1980 Periodic spontaneous collapse and revival in a simple quantum model *Phys. Rev. Lett.* **44** 1323

[2] Weisskopf V and Wigner E 1930 Calculation of the natural brightness of spectral lines on the basis of Dirac's theory *Z. Phys.* **63** 54

IOP Publishing

Chapter 7

Coherence and detection

In quantum mechanics, we do not know anything unless we measure it, and the process of measurement must generally be specified if we wish to predict a result. In this chapter we give some details of how photodetectors work and discuss what they actually measure. The simplest thing to model is a photodetector that absorbs a photon from the field. This could be a two-level atom initially in its ground state, and when it absorbs a photon it is in the ground state. A potential can be applied that is sufficient to ionize the atom in the excited state but not in the ground state, and this then provides a very weak current pulse (due to the one ionized electron) which can then be amplified. Most photodetectors work on some version of this process, to convert a photon (or bunch of photons) into a macroscopic burst of electrons that yields a measurable current pulse. Let us start with a simple version of the measurement process. Assume that we have a photodetector that works by removing a photon from the field and yields a current pulse that is spatially and temporally localized. We will refer to this as the 'click' of the detector, reminiscent of the noise the Geiger counter [7] makes in science fiction movies, as well as real Geiger counters! The effect of the detection is then to transform the state

$$|\psi\rangle_{AD} = a|\psi\rangle_{BD} \tag{7.1}$$

where AD and BD represent after and before detection. Notice that this model of measurement is not a unitary process, and does not preserve the length of the state vector.

It is tempting, based on the relation

$$a|n\rangle = \sqrt{n}|n-1\rangle \tag{7.2}$$

to think that the application of a to a quantum state represents taking away a photon from the field, removing $\hbar\omega$ of energy in the process. The most general state is a superposition of number states. Consider the action of a on a coherent state, recall we have

$$a|\alpha\rangle = |\alpha\rangle \tag{7.3}$$

and does not even change the state, the average photon number and variance are unchanged. As a further example, consider the state

$$|\psi\rangle = \sqrt{1 - \varepsilon^2}|0\rangle + \varepsilon|2\rangle \tag{7.4}$$

which has mean photon number of $\langle n \rangle = 2\varepsilon^2$. When a acts on this state, we find

$$a|\psi\rangle = \sqrt{2}\varepsilon|1\rangle \rightarrow |1\rangle \tag{7.5}$$

where we have normalized the state after the action of a. Here, we see that we have a state with a definite photon number of one. The application of a has actually increased the average photon number! So, one must take care with our interpretation of what a does to a state! We will have more to say about this in chapter 9 when we consider quantum trajectory theory.

7.1 Detection of a noiseless classical signal

Let us assume that we have a classical electromagnetic wave of intensity $\bar{I}$ with no fluctuations and that we have a detector that generates a signal proportional to that intensity. Let us say we have a detector that has a probability of giving a 'click', or photodetection event, over some time interval dt given by $sIdt$. At time $t + dt$, if we have detected m photons, with probability $P_m(t + dt)$, this could have occurred in two ways. We could have detected $m - 1$ photons in the interval $[0, t]$, and one photon in the interval $[t, t + dt]$. It is also possible that one detected m photons in the interval $[0, t]$ and none in the following interval $[t, t + dt]$. This leads us to

$$P_m(t + dt) = P_{m-1}(t)sIdt + P_m(t)(1 - sIdt) \tag{7.6}$$

where the probability of *not* firing is $(1 - sIdt)$. Rearranging a bit we have

$$\frac{P_m(t + dt) - P_m(t)}{dt} = (P_{m-1}(t) - P_m(t))sI \tag{7.7}$$

and take the limit $dt \rightarrow 0$ to find

$$\frac{dP_m}{dt} = sI(P_{m-1}(t) - P_m(t)) \tag{7.8}$$

The solution is

$$P_m(t) = \frac{(sIt)^m}{n!}e^{-sIt} \tag{7.9}$$

where we have used as boundary conditions the fact that the probability of detecting a photon in a time interval of zero length is zero,

$$P_0(t + 0) = 1 \tag{7.10}$$

$$P_m(t + 0) = 0, \quad m \geqslant 0 \tag{7.11}$$

This is a Poisson distribution with average photocount number $\bar{m} = sIt$, and a variance of $\sqrt{\bar{m}}$. This uncertainty, or noise, in the photocount arises from the random nature of the detector firing, and we will come to call this 'shot noise'. This noise occurs for a light source with a constant intensity, and no fluctuations. Hence, this noise is due to the detection process, and later we will find a relation to vacuum fluctuations. If we count photons for a time T, and we allow the intensity to change in time, then the above formulas hold with

$$\bar{I} = \frac{1}{T} \int_0^T I(t) dt \tag{7.12}$$

To obtain the statistical photocount distribution, one generally counts photons for a time T, and then repeats it for many such intervals. This is then given by an ensemble average,

$$P_m(t) = \left\langle \frac{\lambda^m}{m!} e^{-\lambda} \right\rangle \tag{7.13}$$

with $\lambda = s\bar{I}T$, which is known as the Mandel counting formula [9].

Using this distribution we can calculate the mean number of counts in time T,

$$\begin{aligned} \langle m \rangle &= \sum_{m=0} m P_m \\ &= \sum_{m=0} \frac{\lambda^m}{(m-1)!} e^{-\lambda} \\ &= \lambda \sum_{m=1} \frac{\lambda^{m-1}}{(m-1)!} e^{-\lambda} \\ &= \lambda \end{aligned} \tag{7.14}$$

In a similar manner, one can show that

$$\langle m(m-1) \rangle = \lambda(\lambda - 1) \tag{7.15}$$

We can then calculate

$$\begin{aligned} (\Delta m)^2 &= \langle m^2 \rangle - \langle m \rangle^2 \\ &= \lambda \end{aligned} \tag{7.16}$$

which is the usual result for a Poissonian distribution, with the variance equal to the mean. So, we see that for a classical field with no fluctuations at all, there is uncertainty, or noise, in the distribution of counts. For many intervals T long, we will get a range of photocounts on the order of $\lambda - \sqrt{\lambda}$ to $\lambda + \sqrt{\lambda}$, roughly speaking. What is the source of this noise? As there are no fluctuations in the light wave, it would seem that the noise comes from the detection process itself, indeed from the random nature of whether the detector fires. This type of noise is a form of what is referred to as 'shot' noise; we will have more to say when we consider the spectrum of squeezing later. For a field with fluctuations, we have the variance

$$(\Delta m)^2 = \langle \lambda \rangle + (\langle \lambda^2(t, T) \rangle - \langle \lambda \rangle^2) \tag{7.17}$$

where again the first term is the 'shot' noise contribution.

It is often useful to consider a waiting time distribution, that is, what is the distribution of times until the *next* photon is detected. We break up the time interval $T = Ndt$ and consider that in all N of those time intervals there is no photodetection, but in the interval $N + 1$ there is one. Then we have

$$\begin{aligned} w(T)dt &= (1 - s\bar{I}dt)^N s\bar{I}dt \\ &= (1 - s\bar{I}T/N)^N s\bar{I}dt \end{aligned} \tag{7.18}$$

In the limit that $N \to \infty$, we have

$$w(T) = e^{-s\bar{I}T} s\bar{I}T \tag{7.19}$$

Let us now consider a light field in thermal equilibrium at temperature T. The probability of the field to have n photons in it is

$$P_n^{th} = \frac{1}{Z} e^{-E_n/kT} \tag{7.20}$$

where $Z = \sum_n P_n$. Recalling that the energy of the nth energy level is $E_n = (n + 1/2)\hbar\omega$. Ignoring the zero point energy, which cancels in P_n anyway, we can sum

$$Z = \sum_n e^{-n\hbar\omega/kT} = \frac{1}{1 - e^{-\hbar\omega/kT}} \tag{7.21}$$

The average photon number is then

$$\begin{aligned} \langle n \rangle &= \sum_n n P_n \\ &= (1 - e^{-\hbar\omega/kT}) \sum_n n e^{-E_n/kT} \\ &= (1 - e^{-\hbar\omega/kT}) \frac{\partial}{\partial \beta} \sum_n e^{-E_n/kT} \\ &= (1 - e^{-\hbar\omega/kT}) \frac{\partial}{\partial \beta} \frac{(-e^{-\hbar\omega/kT})}{(1 - e^{-\hbar\omega/kT})^2} \\ &= \frac{1}{e^{\hbar\omega/kT} - 1} \end{aligned} \tag{7.22}$$

This allows us to write the photon probability as

$$P_n = \frac{\langle n \rangle^n}{(1 + \langle n \rangle)^{1+n}} \tag{7.23}$$

For large $\langle n \rangle$, this can be approximated as

$$P_n = \frac{1}{\langle n \rangle} e^{-n/\langle n \rangle} \tag{7.24}$$

where this is only really appropriate for large photon numbers and we treat n as a continuous variable. In this regime we can also use the intensity $I \approx n$,

$$P(I) = \frac{1}{\langle I \rangle} e^{-I/\langle I \rangle} \tag{7.25}$$

7.2 Complex analytic signal

To consider the effect of fluctuations on the detected signal, we first consider the Fourier decomposition of the field,

$$E(t) = \frac{1}{2\pi} \int_{-\infty}^{\infty} d\omega \mathcal{E}(\omega) e^{-i\omega t} d\omega \tag{7.26}$$

Generally in our applications the frequency content of the field, as given by $e(\omega)$ consists of frequencies small compared to the central oscillation frequency ω_0. Since the field is a real quantity, $E(t) = E^*(t)$, we have

$$\begin{aligned} \frac{1}{2\pi} \int_{-\infty}^{\infty} d\omega \mathcal{E}(\omega) e^{-i\omega t} &= \frac{1}{2\pi} \int_{-\infty}^{\infty} d\omega \mathcal{E}^*(\omega) e^{i\omega t} \\ &= \frac{1}{2\pi} \int_{-\infty}^{\infty} d\omega \mathcal{E}^*(-\omega) e^{-i\omega t} \end{aligned} \tag{7.27}$$

or more succinctly, ˥$(\omega) = \mathcal{E}^*(-\omega)$. Then we can define

$$E(t) = E^{(+)}(t) + E^{(-)}(t) \tag{7.28}$$

with

$$E^{(+)}(t) = \frac{1}{2\pi} \int_{0}^{\infty} \mathcal{E}(\omega) e^{-i\omega t} d\omega \tag{7.29}$$

and

$$\begin{aligned} E^{(-)}(t) &= \frac{1}{2\pi} \int_{-\infty}^{0} \mathcal{E}(\omega) e^{-i\omega t} d\omega \\ &= \frac{1}{2\pi} \int_{0}^{\infty} \mathcal{E}(-\omega) e^{i\omega t} d\omega \\ &= \frac{1}{2\pi} \int_{0}^{\infty} \mathcal{E}(\omega) e^{i\omega t} d\omega \end{aligned} \tag{7.30}$$

We then have

$$E(t) = 2Re(E^{(+)}(t)) = 2Re(E^{(-)}(t)) \tag{7.31}$$

We usually refer to $E^{(+)}(t)$ as the positive frequency part of the field, and also as the complex analytic signal. This is often useful in that we can consider the field to be decomposed into a series of complex exponentials $e^{-i\omega t}$ for only positive frequencies, and then we merely take the real part at the end of any sequence of linear operations. In the case of a classical wave of unvarying amplitude, $E(t) = E_0 \cos(\omega_0 t + \phi)$, then we have

$$E^{(+)}(t) = \frac{E_0}{2} e^{-i(\omega_0 t + \phi)} \tag{7.32}$$

$$E^{(-)}(t) = \frac{E_0}{2} e^{i(\omega_0 t + \phi)} \tag{7.33}$$

For such a case, the Fourier transform of the field consists of two delta functions, one at $\omega = \omega_0 = \omega$ and another at $\omega = -\omega_0$. In many cases of interest, the bandwidth of the light $\Delta\omega$ is very small compared to the carrier frequency ω_0. In this case, we can approximate the light field by

$$E^{(+)}(t) = \frac{E_0(t)}{2} e^{-i(\omega_0 t + \phi(t))} \tag{7.34}$$

$$E^{(-)}(t) = \frac{E_0(t)}{2} e^{i(\omega_0 t + \phi(t))} \tag{7.35}$$

where $E_0(t)$ and $\phi(t)$ are the slowly varying amplitude and phase of the light field, in the same sense as in our earlier chapters, $\dot{E}_0 \ll \omega_0 E_0$ and $\dot{\phi} \ll \omega_0 \phi$. What about the intensity of the light field? We have

$$\begin{aligned} I &= \varepsilon_0 c |E(t)|^2 \\ &= \varepsilon_o c \frac{E_0^2}{4} (2 + e^{2i(\omega t + \phi)} + e^{-2i(\omega t + \phi)}) \\ &= \varepsilon_0 c \frac{E_0^2}{2} (1 + \cos(2(\omega t + \phi))) \\ &= \varepsilon_0 c E_0^2 \cos^2(\omega t + \phi) \\ &= \frac{\varepsilon_0 c}{4} (|E^{(+)}(t)|^2 + |E^{(-)}(t)|^2 + E^{(+)}(t) E^{(-)}(t) + E^{(+)}(t) E^{(-)}(t) +) \end{aligned} \tag{7.36}$$

Taking a time average, we find

$$\begin{aligned} \bar{I} &= \varepsilon_0 c \frac{E_0^2}{2} \\ &= \frac{\varepsilon_0 c}{4} (E^{(+)}(t) E^{(-)}(t) + E^{(+)}(t) E^{(-)}(t) +) \\ &= \frac{\varepsilon_0 c}{2} E^{(+)}(t) E^{(-)}(t) \end{aligned} \tag{7.37}$$

where we have used the fact that $E^{(+)}(t)E^{(-)}(t) = E^{(-)}(t)E^{(+)}(t)$. When we treat the field quantum mechanically, we will find (actually we already have...) that $E^{(+)}(t)$ and $E^{(-)}(t)$ are noncommuting operators.

7.3 Semiclassical photodetection theory

We must specify how the detector actually works. The model we will consider here is an atom initially in the ground state, that can be excited into an excited state, or a continuum of excited states. At that point a voltage is applied that will ionize the detector atom, and yield a current due to one electron. Then some type of amplification occurs that turns that one electron into a macroscopic burst of electrons. This of course requires energy, and oftentimes the detector must recharge that energy, and the detector cannot 'fire' for some time interval, often referred to as 'dead time'. In this section, we treat the atom quantum mechanically, but not the field. We will treat the field quantum mechanically in the next section.

We assume that there is an electric dipole interaction between the field to be detected and the detector atoms

$$\begin{aligned} H_{int} &= -\vec{\mu} \cdot \vec{E} \\ &= -\mu_{eg}(\sigma_- e^{-i\omega_{if} t} + \sigma_+ e^{i\omega_{if} t}) E_0(t) \cos(\omega_0 t) \\ &= -\frac{\hbar\Omega(t)}{2}(\sigma_- e^{-i(\omega_{if}-\omega_0)t} + \sigma_- e^{i(\omega_{if}-\omega_0)t}) \end{aligned} \tag{7.38}$$

as in chapter 3, although at this point we are in the interaction picture, $\Omega = \mu_{eg} E_0/\hbar$ and $\omega_{if} = \omega_i - \omega_f$ is the frequency difference between the ground (initial) and excited (final) states. Also, we take the incident field to be of the form $E(t) = E_0(t)\cos(\omega_0 t)$. Terms oscillating at $\omega_0 + \omega_{if}$ have been neglected, the rotating wave approximation. The formal solution to the Schrödinger equation in the interaction picture is

$$|\psi(t)\rangle = -\frac{i}{\hbar}\int_0^t dt' H_I(t')|\psi(t')\rangle \tag{7.39}$$

In the first-order perturbation theory, we have

$$|\psi(t)\rangle = -\frac{i}{\hbar}\int_0^t dt' H_I(t')|\psi(0)\rangle \tag{7.40}$$

The probability of finding the system in a final state $|f\rangle$ given that it started in an initial state $|i\rangle$ is given by

$$P_{i\to f} = |\langle f|\psi(t)\rangle|^2 \tag{7.41}$$

which can be written

$$\begin{aligned} P_{i\to f} &= \left| -\frac{i}{\hbar}\int_0^t \langle iH_{int}(t')|f\rangle \right|^2 \\ &= \frac{1}{\hbar^2}\int_0^t dt' \int_0^{t'} dt'' \langle i|H_{int}(t')|f\rangle\langle f|H_{int}(t'')|i\rangle \end{aligned} \tag{7.42}$$

At this point, we assume that there is a range of final states, and calculate

$$P_{\text{click}} = \sum_f P_{i \to f} \tag{7.43}$$

where we use a sum of probabilities. We extend the sum from one over all final states accessible from the initial state, to all states of the atom. This basically adds nothing to the sum, as the states that are not accessible have a zero matrix element between the initial state and the interaction Hamiltonian. We can then use

$$\sum_{\text{allstates}} |f\rangle\langle f| = 1 \tag{7.44}$$

to find

$$P_{\text{click}} = \frac{1}{\hbar^2}\int_0^t dt' \int_0^{t'} dt'' \langle i|H_{int}(t')H_{int}(t'')|i\rangle \tag{7.45}$$

We must also integrate over the density of final states,

$$P_{\text{click}} = \frac{1}{\hbar^2}\int_0^t dt' \int_0^{t'} dt'' \int_{\omega_{if}}^{\infty} d\omega\rho(\omega_{if})\langle i|H_{int}(t')H_{int}(t'')|i\rangle \tag{7.46}$$

where the lower limit is the energy gap between the initial state and the lowest final state and extend the upper limit to ∞. Writing $\cos(\omega_0 t) = (1/2)(\exp(i\omega_0 t) + \exp(-i\omega_0 t))$ and putting in the form of the interaction Hamiltonian, the expectation value in equation (7.55) takes the form

$$\begin{aligned}\langle i|H_{int}(t')H_{int}(t'')|i\rangle = {} & \frac{\Omega(t')\Omega(t'')}{4}\langle i|\sigma_-^2 e^{i(\omega_0-\omega_{if})(t'+t'')} + \sigma_+^2 e^{-(\omega_0-\omega_{if})(t'+t'')} \\ & + \sigma_-\sigma_+ e^{i(\omega_0-\omega_{if})(t'-t'')} + \sigma_+\sigma_- e^{-i(\omega_0-\omega_{if})(t'-t'')}|i\rangle\end{aligned} \tag{7.47}$$

with the initial state being the ground state, and using $\sigma_-^2 = \sigma_+^2 = 0$, we find

$$P_{\text{click}} = \frac{1}{4}\int_0^t dt' \int_0^{t'} dt'' \int_{\omega_{if}}^{\infty} d\omega\Omega(t')\Omega(t'')\rho(\omega_{if})e^{i(\omega_0-\omega_{if})(t'-t'')} \tag{7.48}$$

The density of final states is a slowly varying function compared to the complex exponential, and we pull it out of the frequency integral, which can then be performed via

$$\int_{-\infty}^{\infty} d\omega e^{i(\omega_0-\omega_{if})(t'-t'')} = 2\pi\delta(t' - t'') \tag{7.49}$$

which leaves us with

$$P_{\text{click}} = \frac{\pi\rho(\omega_0)}{2}\int_0^t dt'(\Omega(t'))^2 \tag{7.50}$$

We now perform an ensemble average over and use the complex analytic signal to write

$$P_{\text{click}} = \frac{\pi\rho(\omega_0)\mu_{eg}^2}{2}\int_0^t dt'\langle E^{(-)}(t')E^{(+)}(t')\rangle \tag{7.51}$$

We can define an instantaneous rate via

$$R \equiv \frac{dP_{\text{click}}}{dt} = \frac{\pi\rho(\omega_0)\mu_{eg}^2}{2}\langle E^{(-)}(t)E^{(+)}(t)\rangle \tag{7.52}$$

Please note the ordering of the operators, the annihilation operator is to the right, and the creation is to the left. This is called normal ordering and represents the absorptive nature of the detector considered. Earlier in chapter 3, when we considered the interaction of a two-level atom with a classical field, we also saw a constant excitation rate; in that case this was due to considering shining light of many frequencies on the two-level atom. That resulted in a frequency integral much like the integration over the final atomic states above. There we remarked that a constant result would also result from shining monochromatic light on an ensemble of Doppler or pressure broadened atoms. Here, we see that a constant rate results from shining monochromatic light on an atom with many possible excited states that form a continuum. This is quite similar to the photoelectric effect, and with good reason. It *is* the photoelectric effect! Recall that the photoelectric effect occurs when monochromatic light is incident on a metal; the metal has a filled valence band, and a slightly occupied conduction band. In the photoelectric effect, it was found experimentally that the light could kick out electrons and that the emitted electrons had energy

$$E = \hbar\omega - \Phi \tag{7.53}$$

where ω is the frequency of the emitted light and Φ is the work function of the metal, the energy required to get the electron to the conduction band or ionize it. If the frequency is less than $\Phi/\hbar$ no electrons are emitted. Classically, the energy of the incident light beam is proportional to the intensity. It was puzzling initially that it was the frequency of the light that determined the energy of the ejected electron. Einstein won the Nobel prize, not for relativity, but for his interpretation of the photoelectric effect [1] that the light/matter interaction in this case is such that light exhibits a particle-like behavior. In many introductory 'modern' physics courses, this is presented as evidence for the photon nature of the field. That is, that light is composed of particles of energy $E = \hbar\omega$, which can cause an electron to be emitted if the photon has enough energy to promote the electron above the gap. In this section, we have considered the electric field to be a classical number that oscillates in time. Where are the photons? We have found an excitation rate that is constant, which implies that there can be a zero delay time for electron emission. Also, the frequency of light must be such that $\omega \geqslant \omega_0$. The energy of the emitted electron varies linearly with the frequency of the input light. In short, there are no photons in this, in the sense of a proof that light is composed of particles called photons with energy $E = \hbar\omega$. Does this mean there are no photons? Of course not. All it means is that the photoelectric effect and photodetectors do not provide direct indisputable evidence for photons, but their explanation is certainly consistent with photons.

7.4 Quantum detection theory

Here, we model the interaction of a quantum field with the atoms in our detector. To this end, we utilize the interaction Hamiltonian, in the interaction picture

$$H_{int} = \hbar g(a\sigma_+ e^{-i\Delta t} + a^\dagger \sigma_- e^{-i\Delta t}) \tag{7.54}$$

where we have taken the detector atom to be at an antinode of the field and made the rotating wave approximation. The click rate is then given by

$$P_{\text{click}} = \frac{1}{\hbar^2}\int_0^t dt' \int_0^{t'} dt'' \langle i|H_{int}(t')H_{int}(t'')|i\rangle \tag{7.55}$$

where the $\langle\rangle$ refers to a quantum expectation value in the initial state, which now must exist in a Hilbert space of the atom and field. The quantum treatment follows exactly the same steps as the semiclassical approach, which led to

$$P_{\text{click}} = \frac{\pi\rho(\omega_0)\mu_{eg}^2}{2}\int_0^t dt' E^{(-)}(t')E^{(+)}(t')(\langle g|\sigma_-\sigma_+|g\rangle + \langle g|\sigma_+\sigma_-|g\rangle) \tag{7.56}$$

In this case, the initial state is not $|g\rangle$, the atom(s) in the ground state, but $|\psi_F, g\rangle$, where $|\psi_F\rangle$ is the state of the field, and again the atom is in the ground state. Also, instead of $E^{(-)}(t')E^{(+)}(t')$, we have a factor of $\mathcal{E}_0^2$, and with each σ_- operator is paired an $a^\dagger$, and with each σ_+ is paired an a. These operators must of course appear *inside* the expectation value, which leads to

$$\begin{aligned} P_{\text{click}} = \frac{\pi\rho(\omega_0)\mu_{eg}^2\mathcal{E}_0^2}{2}\int_0^t & dt'(\langle\psi_F, g|a^\dagger\sigma_-\sigma_+ a|\psi_F, g\rangle \\ & + \langle\psi_F, g|a\sigma_+\sigma_- a^\dagger|\psi_F, g\rangle) \end{aligned} \tag{7.57}$$

which gives us

$$P_{\text{click}} = \frac{\pi\rho(\omega_0)\mu_{eg}^2\mathcal{E}_0^2}{2}\int_0^t dt' \langle\psi_F|a^\dagger a|\psi_F\rangle \tag{7.58}$$

Now this is no shock perhaps that the click rate of our detector is proportional to the average photon number of the field, $\langle n\rangle = \langle a^\dagger a\rangle$. *But* this result depends crucially on the fact that the atoms of the detector are in the ground state initially and that the atomic and field raising and lowering operators are paired in the way they are. If we had detectors with atoms initially excited, we would then have a detector click rate proportional to $\langle aa^\dagger\rangle = \langle n + 1\rangle$. Classically, we have the intensity proportional to the square of the field. Quantum mechanically, the field is something like $\mathcal{E}_0(ae^{-i\omega t} + a^\dagger e^{i\omega t})$. Squaring this field, and ignoring rapidly oscillating terms, we find that

$$I \propto E^2 = \mathcal{E}_0^2(aa^\dagger + a^\dagger a) \tag{7.59}$$

We write this as

$$I \propto \mathcal{E}_0^2(:aa^\dagger + a^\dagger a:) = 2\mathcal{E}_0^2(a^\dagger a) \tag{7.60}$$

where we have introduced the notation $:O:$ which we refer to as normal ordering. This means move all a's to the right, and all $a^\dagger$'s to the left, *without* using commutation relations. This is not to say, we are now ignoring commutation relations! It merely means that we take classical expression involving fields and intensities, replace them with the quantum field in terms of raising and lower operators for the field. Then we do not go back through perturbation theory, integrating over densities of states, etc, but merely reorder the raising and lowering operators. It is a shortcut and *not* a mere ignoring of operator ordering!

The next example one could consider is related to the probability of detecting two photons.

$$\begin{aligned} \langle n^2 \rangle &\rightarrow \langle :n^2: \rangle \\ &= \langle :a^\dagger a a^\dagger a: \rangle \\ &= \langle a^{\dagger 2} a^2 \rangle \\ &= \langle n(n-1) \rangle \end{aligned} \tag{7.61}$$

Notice that if the field only has one photon in it, the probability of detecting two is zero. If we did *not* use normal ordering, we would have $\langle n^2 \rangle = 1$. So, the normal ordering procedure seems consistent with the discrete nature of the field, that is the photon concept.

We could have obtained this result by recalling the semiclassical result for the click rate of the detector,

$$R \equiv \frac{dP_{\text{click}}}{dt} = \frac{\pi\rho(\omega_0)\mu_{eg}^2}{2}\langle E^{(-)}(t)E^{(+)}(t)\rangle \tag{7.62}$$

In the quantum theory of the electromagnetic field, we have

$$E^{(+)} = \mathcal{E}_0 a e^{-i\omega t} \tag{7.63}$$

$$E^{(-)} = \mathcal{E}_0 a^\dagger e^{i\omega t} \tag{7.64}$$

In the quantum theory, the order matters and we have discovered that we use normal ordering. In the future, we may take semiclassical results, use the prescription above for negative and positive frequency operators, and then apply normal ordering. For example, the photocount distribution is given by

$$P_m(t) = \frac{\lambda^m}{n!}e^{-\lambda} \tag{7.65}$$

where now we use $\lambda = sT\mathcal{E}_0^2 a^\dagger a$ and averages are to be taken as expectation values using the state of the field. This yields

$$\langle m \rangle = sT\mathcal{E}_0^2 \langle a^\dagger a \rangle \tag{7.66}$$

$$\langle m(m-1) \rangle = s^2 T^2 \mathcal{E}_0^4 \langle :(a^\dagger a)^2: \rangle \tag{7.67}$$

7.5 Optical spectra and first-order coherence

Here, we consider a single field mode, but any real mode will actually have some finite linewidth. A spectrum measurement experiment on the system would be in the form of an intensity measurement, so we calculate the intensity distribution as a function of frequency:

$$I(\omega) = E^*(\omega)E(\omega), \tag{7.68}$$

where the frequency distribution of the field is given by the Fourier transform of the electric field's time evolution

$$E(\omega) = \frac{1}{\sqrt{2\pi}} \int_{-\infty}^{\infty} E(t)e^{i\omega t}dt. \tag{7.69}$$

Using this, we see that equation (7.68) can be written as

$$I(\omega) = \frac{1}{2\pi} \int_{-\infty}^{\infty} dt \int_{-\infty}^{\infty} dt' E(t)E^*(t')e^{i\omega(t-t')} \tag{7.70}$$

$$= \frac{1}{2\pi} \int_{-\infty}^{\infty} dt \int_{-\infty}^{\infty} d\tau E(t)E^*(t+\tau)e^{-i\omega\tau}, \tag{7.71}$$

where $\tau = t' - t$. Assuming that the field is ergodic and that the expectation value is equivalent to a time average, we can further simplify equation (7.70) to achieve the Wiener–Khintchine theorem,

$$I(\omega) = \frac{T}{2\pi} \int_{-\infty}^{\infty} d\tau e^{-i\omega\tau} \langle E(t)E^*(t+\tau) \rangle, \tag{7.72}$$

where we have used

$$\langle E(t)E^*(t+\tau) \rangle = \frac{1}{T} \int_{-T/2}^{T/2} dt E(t)E^*(t+\tau). \tag{7.73}$$

T is the total time over which the field is time-averaged. Ideally, this should be very long to resolve the frequency components of interest; in practice, there are limits, and the uncertainty principle tells us that our spectral resolution in the resulting spectra is $\Delta\omega \sim 1/T$. In order to correctly model the experiment, we must make certain that our field $E(t)$ coincides with the experiment's field signal. In the physical experiment, a detector is used to collect the transmitted photons and is thus an absorptive device. This absorption is correctly modeled in our calculation by replacing E with its negative frequency component $E^{(-)} \propto a$. So, now we can write equation (7.72) in terms of the cavity field's creation and annihilation operators

$$I(\omega) = \frac{T}{2\pi} \int_{-\infty}^{\infty} d\tau e^{-i\omega\tau} \langle a^{\dagger}(t+\tau)a(t)\rangle. \tag{7.74}$$

We have used normal ordering (creation operators to the left, annihilation operators to the right) in the above equation, coinciding with our modeling of the detector as an absorptive device.

How could one measure the two time correlation function above? Consider a double slit with light incident on it. The resultant field at a point behind the screen is a superposition,

$$E(t) = f(E_0(t) + E_0(t+\tau)) \tag{7.75}$$

where f is a constant of proportionality (we ignore diffraction effects for the individual slits) and $\tau = \Delta L/c$, where ΔL is the path difference between the paths to the screen from the two slits. The intensity is then, ignoring constants,

$$I(t) = 2|E_0(t)|^2 + 2Re(E_0^*(t)E_0(t+\tau)) \tag{7.76}$$

Taking an ensemble average, we find

$$\begin{aligned} \langle I\rangle &= 2\langle I_0\rangle + 2Re\langle E_0^*(t)E_0(t+\tau)\rangle \\ &= 2\langle I_0\rangle + 2Re\langle E_0^*(0)E_0(\tau)\rangle \end{aligned} \tag{7.77}$$

We have also invoked stationarity of the field statistics, that is

$$\langle E(t)\rangle = \langle E(t+\tau)\rangle = \langle E(0)\rangle \tag{7.78}$$

$$\langle E^*(t)E(t+\tau)\rangle = \langle E^*(0)E(\tau)\rangle \tag{7.79}$$

The field itself is not assumed to be stationary, but the ensemble averages indicated, and higher order ones, are taken to be stationary. An interference pattern is generally observed on a screen behind the double-slits. The fringe visibility is defined as

$$V = \frac{I_{\max} - I_{\min}}{I_{\max} + I_{\min}} \tag{7.80}$$

Typically, for oscillating fields, the correlation function will also oscillate,

$$G^{(1)}(\tau) \equiv \langle E_0^*(0)E_0(\tau)\rangle = I_0 F(\tau)\cos(\omega\tau) \tag{7.81}$$

where $F(\tau)$ has a maximal value of 1 and we have defined the first-order correlation function $G^{(1)}(\tau)$. The maximum intensity is then $I_0F(\tau)$ and the minimum $-I_0F(\tau)$ which leaves us with a visibility

$$V = F(\tau) \tag{7.82}$$

Consider the normalized correlation function

$$g^{(1)}(\tau) = \frac{\langle E_0^*(0)E_0(\tau)\rangle}{\langle E_0(0)\rangle\langle E_0^*(0)\rangle} \tag{7.83}$$

In the situation we consider here, this yields

$$g^{(1)}(\tau) = F(\tau)e^{i\omega\tau} \tag{7.84}$$

We can write the intensity spectrum as

$$\begin{aligned} I(\omega) &= \frac{T}{2\pi}\langle E_0(0)\rangle\langle E_0^*(0)\rangle Re \int_{-\infty}^{\infty} d\tau e^{-i\omega\tau} g^{(1)}(\tau) \\ &= \frac{T}{2\pi}\langle I\rangle Re \int_{-\infty}^{\infty} d\tau e^{-i\omega\tau} g^{(1)}(\tau) \end{aligned} \tag{7.85}$$

Many times there are inelastic collisions, or other inelastic processes which occur in the source producing the light that leads to changes in the phase of the light but not the intensity. This will occur in a Doppler broadened medium, for example. The ensemble average for such a light source results in

$$\begin{aligned} \langle E^*(t)E(t+\tau)\rangle = |E_0|^2 e^{-i\omega_0\tau} \times \\ \langle (e^{-i\phi_1(t)} + e^{-i\phi_2(t)} + e^{-i\phi_3(t)} + \cdots) \\ (e^{i\phi_1(t+\tau)} + e^{i\phi_2(t+\tau)} + e^{i\phi_3(t+\tau)} + \cdots)\rangle \end{aligned} \tag{7.86}$$

where ϕ_n is the relative phase of the n_{th} member of the ensemble. The terms dealing with $\phi_n(t) - \phi_m(t+\tau) \equiv \Delta\phi$ average to zero, leaving terms of the form

$$\langle e^i \Delta\phi\rangle = \left\langle 1 - \frac{1}{2}\Delta\phi^2 + \frac{1}{24}\Delta\phi^2 \right\rangle + \cdots \tag{7.87}$$

$$= e^{-\langle\Delta\phi^2\rangle} \tag{7.88}$$

The final result for the correlation function is

$$\langle E^*(t)E(t+\tau)\rangle = E_0^2 e^{i\omega_0\tau} e^{-\gamma\tau} \tag{7.89}$$

where $\gamma = 1/T_{\text{coll}} \equiv \langle\Delta\phi^2\rangle$, where T_{coll} is the average time between collisions, or phase disturbing events. In general, this would be replaced by the coherence time of the light source, T_C. In terms of our double slit experiment, the larger the γ is, the faster the fringe visibility will decay with increasing delay time (or path length difference). The intensity spectrum for a field with this first-order correlation function is

$$\begin{aligned} I(\omega) &= \frac{T}{2\pi}\langle I\rangle Re \int_{-\infty}^{\infty} d\tau e^{-i(\omega-\omega_0)\tau} e^{-\gamma\tau} \\ &= \frac{T}{2\pi}\langle I\rangle Re \frac{\gamma - i\Delta}{\gamma^2 + \Delta^2} \\ &= \frac{\gamma T}{2\pi}\langle I\rangle Re \frac{1}{\gamma^2 + \Delta^2} \end{aligned} \tag{7.90}$$

where we define $\Delta = \omega_0 - \omega$, and we have a Lorentzian shaped spectra. In the limit that $\gamma \to 0$, the spectrum becomes a delta function. Going back to the time domain via the Young's double slit experiment, a monochromatic light field has a visibility that is constant; that is it does not change even with large delay times, and we get perfect fringes. Any real light field has a finite bandwidth, and then there *will* be a decay of fringe visibility for larger delay times. In the opposite limit, where $\gamma \to \infty$, the spectrum is broad and flat, and we have what is sometimes called white light; that is equal power at all frequencies. This leads to a complete absence of fringes. This is perfectly consistent with our experience that a narrower linewidth will generally lead to better interference effects. While the correlation function may not always exhibit purely exponential decay, in any real light field there must be a finite correlation time, and for a large time delay we will see no fringes, there is no information on what the field will be at a large delay time given the initial value of the field. This is expressed mathematically by the fact that the ensemble average in the numerator factorizes

$$\langle E^{(-)}(0)E^{(+)}(\tau \to \infty)\rangle \to \langle E^{(-)}(0)\rangle\langle E^{(+)(0)}\rangle \tag{7.91}$$

where we have used stationarity. This results in

$$g^{(1)}(\tau \to \infty) \to 1 \tag{7.92}$$

In terms of positive and negative frequency parts, for a perfectly monochromatic field, one has

$$g^{(1)}(\tau) = \frac{\langle E^{(-)}(0)E^{(+)}(\tau)\rangle}{|\langle E^{(-)}(0)\rangle|^2} \tag{7.93}$$

where we have used $\langle E^{(+)}\rangle = \langle E^{(-)}\rangle^*$. Going to our quantum recipe, we have

$$g^{(1)}(\tau) = \frac{\langle a^\dagger(0)a(\tau)\rangle}{\langle a\rangle\langle a^\dagger\rangle} \tag{7.94}$$

We note that for a coherent state, $|\alpha\rangle$, we have $g^{(1)}(\tau) = 1$.

It is often convenient to write the optical spectrum as a coherent spectrum and an incoherent spectrum

Defining

$$\Delta a = \langle a\rangle + \Delta a \tag{7.95}$$

and similarly for $a^\dagger$ we find that the numerator of $g^{(1)}(\tau)$ becomes

$$\langle a^\dagger(0)a(\tau)\rangle = \langle a^\dagger\rangle\langle a\rangle + \langle \Delta a^\dagger(0)\Delta a(\tau)\rangle \tag{7.96}$$

The first term is a constant and yields a δ function upon taking a Fourier transform. This is referred to as the coherent, or elastic scattering. It is just Rayleigh scattering of the incident light, with no change in energy or frequency. The second term can have a non-trivial time dependence, which can give a result with a width, or even multiple peaks, upon Fourier transformation. It results from pairwise (or higher) scattering, as shown in figure 7.1.

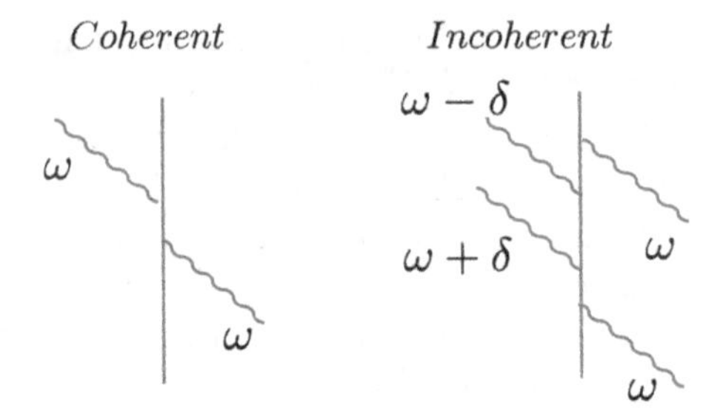

Figure 7.1. Examples of coherent and incoherent scattering processes.

7.6 Photon statistics and second-order coherence

You are told since the earliest discussion of the laser you have in classes that a laser is a bright, monochromatic, directional beam of light. But what about taking a very bright searchlight. It seems fairly directional in nature, it can be quite bright, and if I put a filter on the searchlight that just lets a small range of frequencies through, I have a relatively bright, monochromatic, and directional beam. As a laser is never perfectly monochromatic or directional, and as bright is a meaningless term unless we have something to compare the intensity of the light too, well then how do we distinguish laser light from thermal light from light from some other source? We simply cannot by looking just at the spectrum and mean intensity. We must look at the fluctuations. The simplest thing one can look at perhaps are intensity fluctuations, the variance of the intensity. Consider the decomposition of the intensity into its average value plus noise,

$$I(t) = \langle I(t)\rangle + \Delta I(t) \tag{7.97}$$

Let us now consider a quantity known as

$$g^{(2)}(0) = \frac{\langle I^2\rangle}{\langle I\rangle^2} \tag{7.98}$$

In terms of an average and a fluctuating piece, we have

$$\begin{aligned} g^{(2)}(0) &= \frac{\big\langle\big(\langle I\rangle^2 + (\Delta I)^2 + 2\langle I\rangle\Delta I\big)\big\rangle}{\langle I\rangle^2} \\ &= 1 + \frac{(\Delta I)^2}{\langle I\rangle^2} \end{aligned} \tag{7.99}$$

As the variance of a noise process cannot be negative, we obviously have to have $g^{(2)}(0) \geqslant 1$. For a coherent state, we have

$$g^{(2)}(0) = \frac{|\alpha|^4}{|\alpha|^4} = 1 \tag{7.100}$$

so in some sense we see that the variance of the intensity is zero. Now this is a bit dodgy, as if we look at the 'fuzzball' pictures of a coherent state, there is noise in the electric field amplitude, and one would think that would translate into intensity noise.

Another way to look at this quantity is to write

$$\langle I\rangle(g^{(2)}(0) - 1) = \frac{(\Delta I)^2}{\langle I\rangle} \tag{7.101}$$

that is the difference between $g^{(2)}(0) - 1$ is the difference between variance over the mean and unity, scaled by the average intensity. So, we can see that for a coherent state, $g^{(2)}(0)$ must be 1, as the variance is equal to the mean. The amount of intensity noise in a coherent state is referred to as the standard quantum limit, or shot noise limit. We will have more to say about these terms and $g^{(2)}(0)$ in chapter 6 where we talk about correlation functions and detection schemes. For now we realize that quantum mechanically $I \propto a^\dagger a$, but for the square of the intensity, we must use $I^2 \propto a^{\dagger 2}a^2 \equiv :(a^\dagger a)^2:$ where the notation $:f(a, a^\dagger):$ means to normal order the creation and annihilation operators, move all the creation operators to the left, and all the annihilation operators to the right, *without* using commutation relations. This seems like a procedure gone mad, as the hallmark of quantum mechanics is that order matters in the lab and in our mathematics! The fact that it matters in the lab is what forces us to use mathematics where order matters, like differential operators, matrices, and the like. I am not saying that one just uses any ordering willy-nilly. I am saying (and we will show later) that if detectors work by absorbing photons, as most do, then the normal ordering is the appropriate one. Using this prescription and considering Fock states, we find

$$\begin{aligned} g^{(2)}(0) &= \frac{\langle n|(a^{\dagger 2}a^2)|n\rangle}{\langle n|a^\dagger a|n\rangle^2} \\ &= \frac{\langle n|(a^\dagger a a^\dagger a - a^\dagger a)|n\rangle}{n^2} \\ &= 1 - \frac{1}{n} \end{aligned} \tag{7.102}$$

We see that using this (at presently not totally justified) prescription, we find that while classically the variance may never be less than the mean, for a number state it most certainly is! That is, the noise on the amplitude of the electric field corresponding to a Fock state is less than that of a coherent state. This is a beginning of how we can mathematically calculate properties of our 'quantum fuzzballs' and make connections to measurements. For now, in our next chapter, we consider the interaction of a quantized electric field and a two-level atom. In looking at photon statistics, we are interested in intensity fluctuations of the transmitted field. The question that we ask is, if we detect a photon at some time, then what is the probability of detecting a second photon after a delay time τ. This conditional probability is proportional to

$$G^{(2)}(\tau) = \langle :I(t + \tau)I(t):\rangle \tag{7.103}$$

$$= \langle a^{\dagger}(t)a^{\dagger}(t+\tau)a(t+\tau)a(t)\rangle \tag{7.104}$$

where $I(t) = a^{\dagger}(t)a(t)$ is the intensity operator and as usual $\langle : : \rangle$ denotes normal ordering. Typically, we normalize this probability to the probability for detecting two uncorrelated photons from two uncorrelated photons from a source of the same mean intensity as the source whose statistics we are interested in. This is also the probability of detecting two photons from our source that are separated by an extremely long delay time

$$G^{(2)}(\infty) = \langle a^{\dagger}(0)a(0)\rangle\langle a^{\dagger}(\tau)a(\tau)\rangle \tag{7.105}$$

$$= \langle a^{\dagger}a\rangle^2 \tag{7.106}$$

giving the normalized second-order correlation function

$$g^{(2)}(\tau) = G^{(2)}(\tau)/G^{(2)}(\infty) \tag{7.107}$$

$$= \frac{\langle a^{\dagger}(0)a^{\dagger}(\tau)a(\tau)a(0)\rangle}{\langle a^{\dagger}a\rangle^2}. \tag{7.108}$$

A possible experimental setup for measuring $g^{(2)}(\tau)$ is shown in figure 7.2. One waits for a detection at detector $D2$, waits a time τ and opens a shutter in front of $D1$ and records a detection or not. The second-order intensity correlation function $g^{(2)}(\tau)$ is the probability of two detections τ apart with no information on any other detections. Experimentally, it is much more convenient to let both detectors run and measure the time in between counts. No good experimentalist throws out signal! Using a time division correlator, you can bin those times. This is called the waiting time distribution. It deals with the time until the *next* detection. For very weak fluxes the two are equivalent for several decay lifetimes of the source. Eventually $g^{(2)}(\tau)$ goes to one after many decay times as emissions separated by such a long delay are independent of one another, and the numerator factorizes. For long times the waiting time distribution must go to zero, as there is no chance you have to wait until $t \to \infty$ to detect the *next* photon.

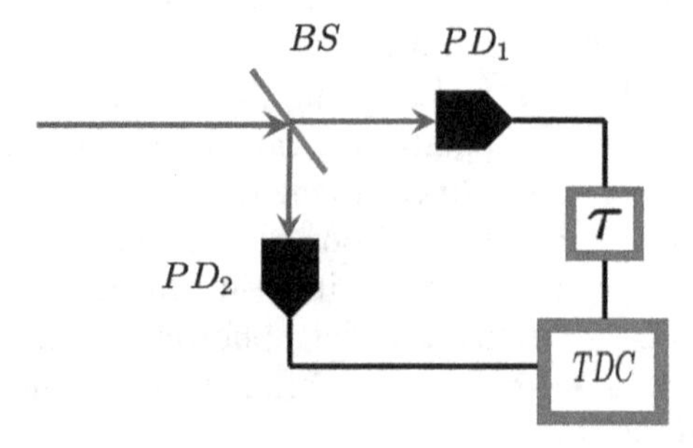

Figure 7.2. Experimental setup for measurement of the waiting time distribution, which is equivalent to $g^{(2)}(\tau)$ for weak fluxes.

Let us begin by considering the zero time correlation,

$$g^{(2)}(0) = \frac{\langle a^{\dagger 2} a^2 \rangle}{\langle a^\dagger a \rangle^2} \tag{7.109}$$

Using commutation relations, we find that this can be written in the form

$$\begin{aligned} g^{(2)}(0) &= \frac{\langle n(n-1) \rangle}{\langle n \rangle^2} \\ &= \frac{(\Delta n)^2}{\langle n \rangle^2} + 1 - \frac{1}{\langle n \rangle} \\ &= \frac{1}{\langle n \rangle}(F - 1) + 1 \end{aligned} \tag{7.110}$$

where we have defined the Fano factor [5] as the variance over the mean,

$$F = \frac{(\Delta n)^2}{\langle n \rangle} \tag{7.111}$$

A slightly related form is

$$\langle n \rangle (g^{(2)}(0) - 1) = F - 1 \tag{7.112}$$

A related quantity is the Mandel Q parameter [10]

$$Q = F - 1 \tag{7.113}$$

Classically, we would have

$$g^{(2)}(0) = \frac{\langle I^2 \rangle}{\langle I \rangle^2} \tag{7.114}$$

Writing $I = \langle I \rangle + \Delta I$, we find,

$$\begin{aligned} g^{(2)}(0) &= \frac{\langle (\langle I \rangle + \Delta I)^2 \rangle}{\langle I \rangle^2} \\ &= \frac{\langle I \rangle^2 + \langle (\Delta I)^2 \rangle + 2\langle I \rangle \langle \Delta I \rangle}{\langle I \rangle^2} \\ &= \frac{\langle (\Delta I)^2 \rangle}{\langle I \rangle^2} + 1 \end{aligned} \tag{7.115}$$

For a classical field which has a well defined value at any instant in time, even if it is noisy, we must have $(\Delta I)^2 \geqslant 0$ which leads to an inequality that must be satisfied by a classical field

$$g^{(2)}(0) \geqslant 1 \tag{7.116}$$

When $g^{(2)}(0) = 1$, this denotes Poissonian statistics as from a coherent state that the variance equals the mean. When $g^{(2)}(0) > 1$, this indicates that pairs of photons are likely to be detected and is called photon bunching. In the case $g^{(2)}(0) < 1$ indicates that photons are not likely to come in pairs and that this results in photon antibunching, or amplitude squeezed light, with reduced intensity fluctuations.

How can this be? How can I have a stochastic variable, say $E(t)$ or $I(t)$, that has a negative variance? The short answer is that we *cannot*! Finding a field which has $g^{(2)}(0) < 1$ would mean finding one which has no real stochastic process associated with it. This means that we cannot consider such a field to have a well defined field amplitude or intensity, even if noisy. This means that objective reality does not exist independent of our interaction with it. There is no wiggly noisy field that exists without our measuring it. This is a key part of quantum weirdness that quantum mechanics is more than just statistical in nature, like the velocities of atoms in a gas in a room. This is no weirder than realizing that the electron in a hydrogen atom has no well defined position without our observing it at a definite position. In a later chapter we will talk about other types of quantum weirdness to be sure. Do such nonclassical states exist? Sure! First, consider a Fock state, $|n\rangle$, which has

$$g^{(2)}(0) = \frac{n(n-1)}{n^2} = 1 - \frac{1}{n} \tag{7.117}$$

Any Fock state other than the vacuum violates the classical inequality, although the violation gets smaller for larger and larger values of n. A coherent state would yield $g^{(2)}(0) = 1$, and is second-order coherent as well as first-order coherent. In fact, we will find that any normalized correlation function for a coherent state is unity. For thermal light, we can show (**do this**) that

$$g^{(2)}(0) = 1 + |g^{(0)}(0)|^2 = 2 \tag{7.118}$$

and for squeezed states, we find

$$g^{(2)}(0) = 1 + \frac{e^{-2r} - 1}{\sinh^2(r)} \tag{7.119}$$

For resonance fluorescence, where a single two-level atom is illuminated by a monochromatic field, we find $g^{(2)}(0) = 0$ as it is impossible for a single atom to emit two fluorescent photons simultaneously, the atom must absorb in between the emission events. For N atoms, one finds that $g^{(2)}(0) = 1 - 1/N$, and the antibunching effect vanishes for a large ensemble of atoms. A violation of the inequality equation (7.116) is referred to as photon antibunching as the photons tend not to come in pairs. Also, it means that the variance is less than the mean, which means the intensity fluctuations are sub-Poissonian.

The classically allowed ranges of the Mandel Q parameter are

$$Q \geqslant 0 \tag{7.120}$$

Mandel Q parameters in the range $-1 \leqslant Q < 0$ would imply a nonclassical field, similarly for F, a negative value implies nonclassicality.

The time dependence of $g^{(2)}(\tau)$ can also exhibit nonclassical behavior. Using the Cauchy–Schwarz inequality one can show that for a classical field we have

$$g^{(2)}(\tau) \leqslant g^{(2)}(0) \tag{7.121}$$

This can be shown by considering a joint probability distribution function $P(I_0, 0; I_{\tau,\tau})$, which is the probability of finding an intensity $I(\tau)$ at time τ given that one had intensity $I(0)$ at $\tau = 0$. We have

$$\langle I(\tau)I(0)\rangle = \int_0^\infty dI_\tau \int_0^\infty dI_0 I_\tau I_0 P(I_0, 0; I_\tau, \tau) \tag{7.122}$$

and

$$\langle I\rangle = \int_0^\infty dI I P(I) \tag{7.123}$$

where $P(I)$ is the probability of the intensity, a single time stationary probability. Using the Cauchy–Schwarz inequality in the form

$$\left|\int f(x)g(x)dx\right|^2 \leqslant \left|\int f(x)dx\right| \left|\int g(x)dx\right| \tag{7.124}$$

we can show that

$$\langle I(\tau)I(0)\rangle \leqslant \langle I^2(0)\rangle \tag{7.125}$$

which leads to equation (7.121). A violation of this inequality was first observed in resonance fluorescence in an atomic beam by Kimble, Dagenais, and Mandel [8], and more recently for a single atom in a trap by the group of Walther. For historical reasons, a violation of this inequality is also referred to as photon antibunching.

We can also show that

$$|g^{(2)}(\tau) - 1| \leqslant |g^{(2)}(0) - 1| \tag{7.126}$$

This can be shown by considering $I(t) = \langle I\rangle + \Delta I(t)$. We can write

$$g^{(2)}(\tau) = 1 + \frac{\langle \Delta I(\tau)\Delta I(0)\rangle}{\langle I\rangle^2} \tag{7.127}$$

and using the Cauchy–Schwarz inequality above, but this time in terms of $\Delta I(\tau)$, resulting in

$$|\langle \Delta I(\tau)\Delta I(0)\rangle| \leqslant \langle(\Delta I^2)\rangle \tag{7.128}$$

which leads to equation (7.126).

7.7 Balanced homodyne detection and the spectrum of squeezing

Consider the setup in figure 7.3.

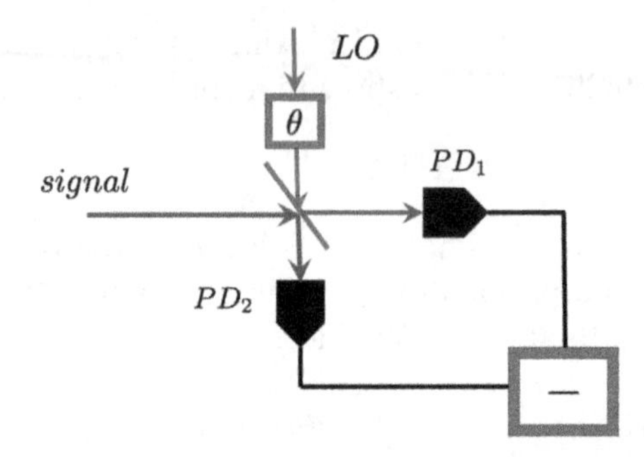

Figure 7.3. Experimental setup for the measurement of the spectrum of squeezing. The difference current is Fourier transformed to obtain the result.

We have a signal field and a local oscillator field incident on a beam splitter. We describe the signal field by a, and the local oscillator field by b. If the beam splitter has $r = t = 1/\sqrt{2}$, the two output modes of the beamsplitter are

$$c = \frac{1}{\sqrt{2}}(a + b) \tag{7.129}$$

$$d = \frac{1}{\sqrt{2}}(a - b) \tag{7.130}$$

The photon numbers in the two output modes are given by

$$n_c = \frac{1}{2}(a^\dagger a + b^\dagger b + ab^\dagger + ba^\dagger) \tag{7.131}$$

$$n_d = \frac{1}{2}(a^\dagger a + b^\dagger b - ab^\dagger - ba^\dagger) \tag{7.132}$$

The difference between these two photon numbers is

$$n_c - n_d = (ab^\dagger + ba^\dagger) \tag{7.133}$$

and the difference between the two detector rates is proportional to

$$\langle n_c \rangle - \langle n_d \rangle = \langle (ab^\dagger + ba^\dagger) \rangle \tag{7.134}$$

If the local oscillator is in a coherent state $\alpha e^{i\phi}$, we then have

$$\langle n_c \rangle - \langle n_d \rangle = \alpha \langle (ae^{-i\phi} + e^{i\phi} a^\dagger) \rangle \tag{7.135}$$

where we recognize this as the expectation value of a field quadrature operator,

$$\langle n_c \rangle - \langle n_d \rangle = 2\alpha \langle a_\theta \rangle \tag{7.136}$$

The variance of the difference in photocurrents is

$$\langle (\Delta n)^2 \rangle = 4|\alpha|^2 \langle (\Delta a_\theta)^2 \rangle \tag{7.137}$$

Changing the phase of the local oscillator, one can measure the variance of thc field quadratures, and for a squeezed state we can find that the variance falls below the shot noise limit for the quiet quadrature, but of course must be above the shot noise limit for the noisier quadrature.

Any real source of light has a finite bandwidth, and as such has some temporal dynamics. The above variance is a DC variance, and there are many sources of noise that can swamp the quantum noise. It is more useful to consider the correlation of the photocurrent difference $i = i_c - i_d$,

$$\langle i(t)i(t+\tau)\rangle \propto \langle \Delta a_\theta(0)\Delta a_\theta(\tau)\rangle \tag{7.138}$$

and Fourier transform it to obtain a noise spectrum, which we refer to as a spectrum of squeezing,

$$S_\theta(\omega) = \int_{-\infty}^{\infty} d\omega e^{i\omega\tau}\langle \Delta a_\theta(0)\Delta a_\theta(\tau)\rangle \tag{7.139}$$

There is a relation between the incoherent spectrum and a sum of two spectra of squeezing $\pi/2$ out of phase. The presence of squeezing can cause narrowing and other features in the incoherent spectra,

$$I_{inc}(\omega) = S_{\theta,\omega} + S_{\theta+\pi/2,\omega} \tag{7.140}$$

where we have suppressed an overall scale factor.

7.8 Wave–particle duality and conditioned homodyne detection

So far, we have looked at field–field correlations, $g^{(1)}(\tau)$, and intensity–intensity correlations, $g^{(2)}(\tau)$. In this section, we consider a normalized correlation, a field–intensity correlation introduced by Carmichael *et al* [2]

$$h(\tau) = \frac{\langle I(0)E(t)\rangle}{\langle I(0)\rangle\langle E(0)\rangle} \tag{7.141}$$

classically, and a quadrature dependent quantum version

$$h_\theta(\tau) = \frac{\langle a^\dagger(0)a_\theta(\tau)a(0)\rangle}{\langle a^\dagger a\rangle\langle a_\theta\rangle} \tag{7.142}$$

This is not well defined for a light source with a zero mean field for a particular quadrature, but in that event one may add a local oscillator to the source and adjust the size of that local oscillator to maximize the signal-to-noise ratio [4]. This correlation function can be measured by performing a balanced homodyne detection of the field, and only recording data when there has been a photodetection at $\tau = 0$. As such it is a balanced homodyne measurement *conditioned* on detection of a photon at time $\tau = 0$. This can be accomplished by the experimental setup in figure 7.4.

With the added assumption of Gaussian statistics, that allows us to ignore third-order moments, we may then write the Fourier pair:

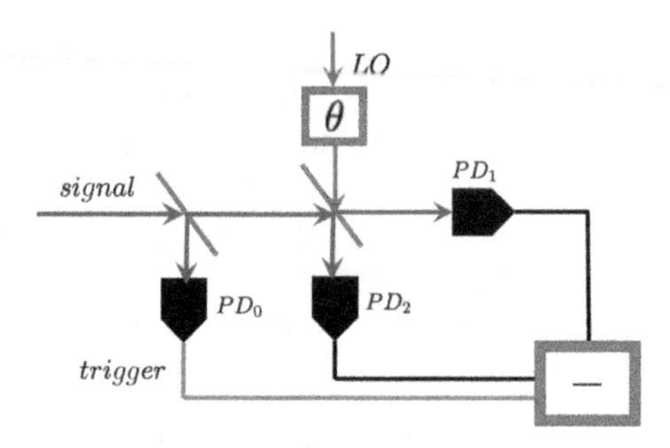

Figure 7.4. Experimental setup for measuring the conditioned balanced homodyne signal $h_\theta(\tau)$.

$$
\begin{aligned}
S(\Omega, \theta) &= 2F \int_{-\infty}^{\infty} d\tau \exp(i2\pi\Omega\tau)[\bar{h}_\theta(\tau) - 1], \\
\bar{h}_\theta(\tau) - 1 &= \frac{1}{4\pi F} \int_{-\infty}^{\infty} d\Omega \exp(-i2\pi\Omega\tau) S(\Omega, \theta).
\end{aligned}
\tag{7.143}
$$

Notice that the photon flux plays a role, in inverse relationship, in the relative sizes of the spectrum of squeezing and the intensity-field correlation function. From this, it would seem that for large photon flux, nonclassical effects might be observed more readily in measurements of the spectrum of squeezing, and for low photon flux, in measurements of $h_\theta(\tau)$. There is also a relationship between the time averaged $\bar{h}_\theta(\tau)$ and the degree of squeezing at zero frequency, and between the frequency averaged spectrum of squeezing and $\bar{h}_\theta(0)$ [11]:

$$
\begin{aligned}
S(0, \theta) &= 2F \int_{-\infty}^{\infty} d\tau[\bar{h}_\theta(\tau) - 1], \\
\bar{h}_\theta(0) - 1 &= \frac{1}{4\pi F} \int_{-\infty}^{\infty} d\Omega S(\Omega, \theta).
\end{aligned}
\tag{7.144}
$$

Also in this limit of Gaussian statistics, one can show the following classical bounds that $h_\theta(\tau)$ must satisfy.

Nonclassical violation of these relations was first observed by Carmichael *et al* [3, 6]

$$
0 \leqslant h_\theta(0) - 1 \leqslant \frac{2}{1 + |\alpha|^2/\langle \Delta a^{\dagger \Delta a} \rangle}
\tag{7.145}
$$

and

$$
|h_\theta(0) - 1| \leqslant |h_\theta(\tau) - 1|
\tag{7.146}
$$

The correlation function $h_\theta(\tau)$ need not be symmetric around $\tau = 0$, in fact an asymmetry is an indication of a breakdown of detailed balance [4].

7.9 Cross-correlation functions

It is also possible, and interesting, to examine correlations between two field modes a and b, the transmission from a cavity and fluorescence out the side of the cavity,

$$g_{ab}^{(2)} = \frac{\langle a^{\dagger}(0)b^{\dagger}(\tau)b(\tau)a(0)\rangle}{\langle a^{\dagger}a\rangle_{SS}\langle b^{\dagger}b\rangle_{SS}} \tag{7.147}$$

A schematic of the measurement apparatus is exhibited in figure 7.5.

The inequalities that this cross correlation satisfy for classical fields are quite different, and are

$$g_{ab}^{(2)}(0) \leqslant \sqrt{g_{bb}^{(2)}(0)g_{aa}^{(2)}(0)} \tag{7.148}$$

$$g_{ab}^{(2)}(\tau) - 1 \leqslant \sqrt{(g_{bb}^{(2)}(0) - 1)(g_{aa}^{(2)}(0) - 1)} \tag{7.149}$$

We can also condition the homodyne detection of mode b with a detection in mode a (figure 7.6),

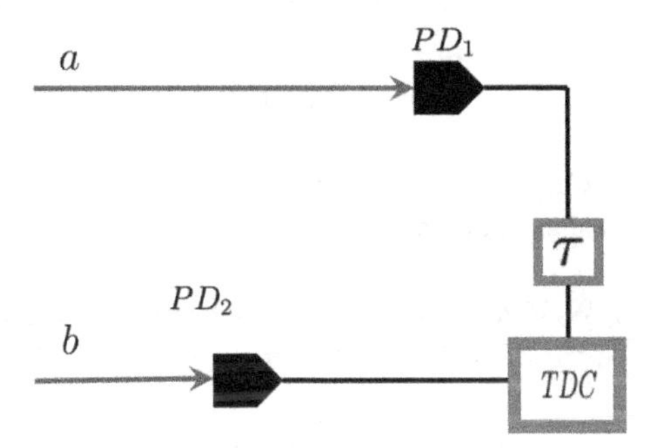

Figure 7.5. Schematic of a measurement of a cross-intensity correlation between field modes a and b.

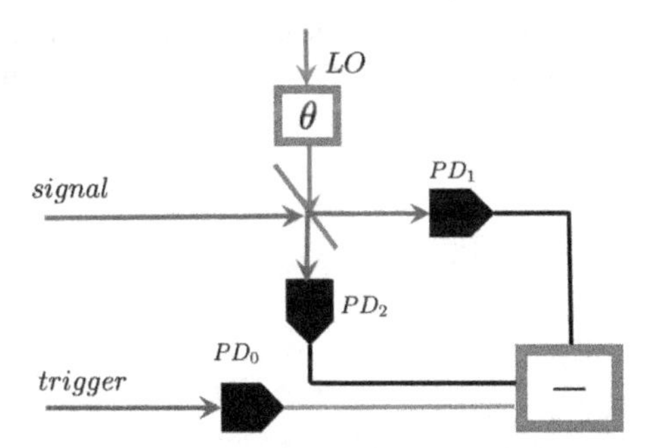

Figure 7.6. Schematic of a measurement of a homodyne detection of one field mode, conditioned by a detection in a different field mode.

$$h_{ab}^{\theta} = \frac{\langle a^{\dagger}(0)b_{\theta}(\tau)a(0)\rangle}{\langle a^{\dagger}a\rangle_{SS}\langle b_{\theta}\rangle_{SS}} \tag{7.150}$$

The inequalities satisfied by this cross h_{θ} are

$$h_{ab}^{\theta}(0) \leqslant \sqrt{h_{bb}^{\theta}(0)h_{aa}^{\theta}(0)} \tag{7.151}$$

$$h_{ab}^{\theta}(\tau) - 1 \leqslant \sqrt{(h_{bb}^{\theta}(0) - 1)(h_{aa}^{\theta}(0) - 1)} \tag{7.152}$$

References

[1] Einstein's nobel prize lecture.

[2] Foster G T, Orozco L A, Castro-Beltran H M and Carmichael H J 2000 Quantum state reduction and conditional time evolution of wave-particle correlations in cavity QED *Phys. Rev. Lett.* **85** 3149–52

[3] Carmichael H, Castro-Beltran H, Foster G and Orozco L A 2000 Giant violations of classical inequalities through conditional homodyne detection of the quadrature amplitudes of light *Phys. Rev. Lett.* **85** 1855–8

[4] Carmichael H J, Foster G T, Orozco L A, Reiner J E and Rice P R 2004 *Intensity-Field Correlations of Non-Classical Light Progress in Optics* vol 46 (Amsterdam: North-Holland)

[5] Fano U 1947 Ionization yield of radiations. II. The fluctuations of the number of ions *Phys. Rev.* **72** 26

[6] Foster G, Orozco L A, Castro-Beltran H and Carmichael H 2000 Quantum state reduction and conditional time evolution of wave-particle correlations in cavity QED *Phys. Rev. Lett.* **85** 3149–52

[7] Geiger H and Müller W 1928 'elektronenzählrohr zur messung schwächster aktivitäten' (electron counting tube for the measurement of the weakest radioactivities) *Die Naturwiss (Sciences)* **16** 617

[8] Kimble H J, Dagenais M and Mandel L 1977 Photon antibunching in resonance fluorescence *Phys. Rev. Lett.* **39** 691

[9] Mandel L 1959 Fluctuations of photon beams: the distributions of the photoelectrons *Proc. Phys. Soc.* **74** 233

[10] Mandel L 1979 Sub-poissonian photon statistics in resonance fluorescence *Opt. Lett.* **4** 205

[11] Reiner J, Smith W P, Orozco L A, Carmichael H and Rice P 2000 Time evolution and squeezing of the field amplitude in cavity QED *J. Opt. Soc. Am.* B **18** 12

IOP Publishing

An Introduction to Quantum Optics

An open systems approach

Perry Rice

Chapter 8

The density matrix and the master equation or wave functions: the big lie

8.1 Open quantum systems

In the 'real world', quantum systems are not isolated from their environment. This is not a lack, as there are systems that we want to interact with by putting energy and information in, and getting energy and information out. This is true in classical mechanics as well, but there we can add driving forces and treat dissipation in an ad hoc phenomenological way.

In classical mechanics, the addition of a velocity dependent force $\vec{F} = -\gamma\vec{v} = -\gamma\frac{\vec{p}}{m}$ can account for dissipation to the environment, as seen in a ballistics problem that accounts for air resistance. Let us consider a similar treatment in quantum mechanics, for adding a damping mechanism to a harmonic oscillator

$$H = \frac{\hat{p}^2}{2m} + \frac{1}{2}m\omega^2\hat{q}^2, \tag{8.1}$$

The equations of motion associated with equation (8.1) after a damping force of $-\gamma p$ is added are

$$\dot{q} = \frac{p}{m}, \qquad \dot{p} = -\gamma p - m\omega^2 q. \tag{8.2}$$

If we are to look at the time evolution of the commutator $[\hat{q}, \hat{p}]$, we see

$$\frac{d}{dt}[\hat{q}, \hat{p}] = \dot{\hat{q}}\hat{p} + \hat{q}\dot{\hat{p}} - \dot{\hat{p}}\hat{q} - \hat{p}\dot{\hat{q}} \tag{8.3}$$

$$= -\gamma[\hat{q}, \hat{p}], \tag{8.4}$$

which has the solution

$$[\hat{q}(t), \hat{p}(t)] = e^{-\gamma t}[\hat{q}(0), \hat{p}(0)] = e^{-\gamma t} i\hbar. \tag{8.5}$$

This obviously cannot be the correct quantum mechanical treatment of dissipation, as the commutator of $\hat{q}$ and $\hat{p}$ would decay with time, making the lower bound of the uncertainty principle decay to 0,

$$\begin{aligned} \Delta q \Delta p &\geqslant \left| \frac{i}{2} \langle [q, p] \rangle \right| \\ &\geqslant \frac{\hbar}{2} e^{-\gamma t}. \end{aligned} \tag{8.6}$$

Because of problems like these in the quantum mechanical treatment of dissipation, the description of a typical quantum optics system that incorporates losses must be fundamentally different than the classical description.

8.2 Density matrix and reduced density matrix

In elementary quantum mechanics we deal with isolated, closed systems. The total energy of the system is conserved, and the system is described by a wavefunction or state vector which contains information about the pure state of the system, including any superpositions or entanglements. A Hermitian Hamiltonian describes the time evolution of the state vector as well as the energy eigenstates of the system. This evolution preserves the norm of the state vector. Quantum systems that interact with their environment cannot generally be described by a wave function, or state vector; we must use a quantity known as a density matrix that characterizes our state of knowledge of the system.

Quantum optics, as a field, began with the invention of the laser, an open system by its nature. Treating open systems complicates the formalism which we use for modeling the system. In quantum optics, we define a system about whose dynamics we wish to know in detail, and an environment about whose dynamics we do not care in detail. We just need gross properties of the environment like temperature, pressure, etc. However, when we allow the system (say a two-level atom) to couple with the environment (all field modes surrounding the atom), then the system is no longer in a pure state, but in a statistical mixture of states. We wish to follow the state of the atom as it absorbs and emits light, but we do not care with which particular mode the atom is interacting. The density operator is the object which describes the statistical mixture of states of an open quantum system. The density operator exists as an object unto itself in a mathematical sense just as the state vector does. It is defined as the sum of projectors onto eigenstates weighted by the probability of being projected onto that state:

$$\rho = \sum_i P_i |\psi_i\rangle\langle\psi_i|. \tag{8.7}$$

However, we can find a matrix representation for the density operator in some basis, which we will call the density matrix. As with any operator, we find a matrix representation by allowing the operator to act on basis vectors. Note that this is not a unique expansion, as any basis may be used. For example, the density matrix in the energy eigenbasis of a two-level atom is

$$\rho \rightarrow \begin{pmatrix} \langle e|\rho|e\rangle & \langle e|\rho|g\rangle \\ \langle g|\rho|e\rangle & \langle g|\rho|g\rangle \end{pmatrix} \tag{8.8}$$

where $|e\rangle$ denotes the excited state and $|g\rangle$ denotes the ground state. To learn something about the properties of a density matrix, let us consider the density matrix for a two level atom in an arbitrary superposition of excited state and ground state

$$|\psi_S\rangle = C_e|e\rangle + C_g|g\rangle \tag{8.9}$$

For the pure state superposition, the density matrix is

$$\begin{pmatrix} \langle e|\psi_S\rangle\langle\psi_S|e\rangle & \langle e|\psi_S\rangle\langle\psi_S|g\rangle \\ \langle g|\psi_S\rangle\langle\psi_S|e\rangle & \langle g|\psi_S\rangle\langle\psi_S|g\rangle \end{pmatrix} = \begin{pmatrix} C_e C_e^* & C_e C_g^* \\ C_g C_e^* & C_g C_g^* \end{pmatrix} \tag{8.10}$$

We see that the diagonal elements of the density matrix tell us populations of states $P_g = |C_g|^2$, $P_e = |C_g|^2$, and the off-diagonal elements tell us about coherence between states. Nonzero off-diagonal elements tell us that in the basis in which we represent the density matrix we have a superposition, which can lead to interference terms in certain measurements. In the state above, we have more than just statistical uncertainty in terms of which state the system is in, excited or ground. The system is in a very real sense in both states simultaneously, which results in quantum interference.

The trace of ρ is just the sum of the probabilities for being in the various states, so normalization of the quantum state results in

$$\mathrm{Tr}(\rho) = 1 \tag{8.11}$$

In the pure state case, as the density operator is really a projection operator onto the state of the system, leaving it in that state, we have

$$\begin{aligned} \rho^2 &= |\psi\rangle\langle\psi|\psi\rangle\langle\psi| \\ &= |\psi\rangle\langle\psi| \\ &= \rho \end{aligned} \tag{8.12}$$

We can also have mixed states more generally, $\rho = \sum_i P_i|\psi_i\rangle\langle\psi_i|$. Here, we have some classical uncertainty in terms of our knowledge of what the quantum state of the system is. For a mixed state which is in the excited state with some probability P_e and the ground state with some probability P_g, the density matrix is

$$\begin{pmatrix} \langle e|\big(P_e|e\rangle\langle e| + P_g|g\rangle\langle g|\big)|e\rangle & \langle e|\big(P_e|e\rangle\langle e| + P_g|g\rangle\langle g|\big)|g\rangle \\ \langle g|\big(P_e|e\rangle\langle e| + P_g|g\rangle\langle g|\big)|e\rangle & \langle g|\big(P_e|e\rangle\langle e| + P_g|g\rangle\langle g|\big)|g\rangle \end{pmatrix} \tag{8.13}$$

which reduces to

$$\begin{pmatrix} P_e & 0 \\ 0 & P_g \end{pmatrix}. \tag{8.14}$$

Here, we *have* a case where either the atom is in the excited or ground state, with some probability, with no interference; just classical statistical uncertainty. For a mixed state, we still have $\mathrm{Tr}(\rho) = 1$ as the probabilities sum to one, but

$$\begin{aligned} \rho^2 &= \sum_{i,j} P_i P_j |\psi_i\rangle\langle\psi_i|\psi_j\rangle\langle\psi_j| \\ &= \sum_j P_j^2 |\psi_j\rangle\langle\psi_j| \\ &\neq \rho \end{aligned} \tag{8.15}$$

One can also show that for a mixed state, we have

$$\mathrm{Tr}(\rho^2) = \sum_j P_j^2 \leqslant \sum_j P_j \tag{8.16}$$

and so $\mathrm{Tr}(\rho^2) \leqslant 1$, where the equality holds for a pure state only. One can use $1 - \mathrm{Tr}(\rho^2)$ as a measure of how far your state is from a pure state; in fact, this is referred to as the impurity of the state.

The decomposition of a mixed state density matrix into a weighted sum of pure state density matrices is not unique. Consider the state

$$\rho = \begin{pmatrix} 1/2 & 1/2 \\ 1/2 & 1/2 \end{pmatrix} \tag{8.17}$$

This can be written as both

$$\begin{aligned} \rho &= 1/2((|e\rangle + |g\rangle(\langle e| + \langle g|)) \\ &= 1/2(|+\rangle + |-\rangle)(\langle +| + \langle -|)) \end{aligned} \tag{8.18}$$

where the dressed states are $|\pm\rangle = (1/\sqrt{2})(|e\rangle \pm |g\rangle)$. In both cases, there is a probability of 1/2 for the two component states.

We can use the density operator to gain information about the system of interest by calculating expectation values and correlation functions of various orders. Expectation values of operators are calculated by taking the trace of both the operator and density matrix over the system states

$$\langle \hat{O} \rangle = Tr_S(\rho \hat{O}) \tag{8.19}$$

which reduces to the standard form

$$\begin{aligned} \langle \hat{O} \rangle &= Tr_S(|\psi\rangle\langle\psi|\hat{O}) \\ &= Tr_S(\langle\psi|\hat{O}|\psi\rangle) \\ &= \langle\psi|\hat{O}|\psi\rangle \end{aligned} \tag{8.20}$$

when $\rho = |\psi\rangle\langle\psi|$ describes a pure state.

In the mixed state case, we also get something that makes sense

$$\begin{aligned}\langle\hat{O}\rangle &= Tr_S\left(\sum_i P_i|\psi_i\rangle\langle\psi_i|\hat{O}\right)\\ &= \sum_i Tr_S(P_i\langle\psi_i|\hat{O}|\psi_i\rangle)\\ &= \sum_i P_i\langle\psi_i|\hat{O}|\psi\rangle\\ &= \sum_i P_i\langle\hat{O}\rangle_i\end{aligned} \tag{8.21}$$

The time evolution of the density matrix under interactions which are described by a Hamiltonian is found by differentiating equation (8.7) and using the Schrödinger equation

$$\dot{\rho} = \frac{d}{dt}\sum_i P_i|\psi_i\rangle\langle\psi_i| \tag{8.22}$$

$$= \sum_i P_i(|\psi_i\rangle\langle\dot{\psi}_i| + |\dot{\psi}_i\rangle\langle\psi_i|) \tag{8.23}$$

$$= -\frac{i}{\hbar}\sum_i P_i(H|\psi_i\rangle\langle\psi_i| - |\psi_i\rangle\langle\psi_i|H) \tag{8.24}$$

$$= -\frac{i}{\hbar}[H, \rho] \tag{8.25}$$

We can now treat systems in mixed states under Hamiltonian evolution using this formalism. However, we wish to treat open quantum systems with dissipation into an environment using the density matrix. For this, we will use a master equation to describe the time evolution of the density matrix.

Although quantum optics generally is not concerned with the specific dynamics of the full environment (we may not be interested in which field mode the atom has radiated into, we care only that the atom has radiated), it is instructive to first characterize a general density operator that represents the state of both the system we are concerned with and the environment. As we have seen, the density operator for a mixed state is the sum of density operators for the possible pure states of the full system (atom + field) with a weight factor of P_i, the probability of being projected onto state i

$$\chi = \sum_i P_i|\psi_i\rangle\langle\psi_i|. \tag{8.26}$$

For a pure state,

$$\chi = |\psi\rangle\langle\psi|. \tag{8.27}$$

This density operator has a matrix representation using the basis vectors of the full system, we call this the density matrix. If our system S (the atom) and the reservoir R (the EM field modes) are coupled together, they form a direct product space $S \otimes R$. Because the number of possible states for the system and the reservoir are different, we designate S as an $N \times N$ space, and R as a $M \times M$ space, with $N \ll M$. The density matrix for $S \otimes R$, $\chi(t)$, takes the form of an $NM \times NM$ matrix,

$$\chi(t) = \begin{pmatrix} R_{11}\rho(t) & R_{12}\rho(t) & R_{13}\rho(t) & \cdots & R_{1M}\rho(t) \\ R_{21}\rho(t) & R_{22}\rho(t) & R_{23}\rho(t) & \cdots & R_{2M}\rho(t) \\ R_{31}\rho(t) & R_{32}\rho(t) & R_{33}\rho(t) & \cdots & R_{3M}\rho(t) \\ \vdots & \vdots & \vdots & \vdots & \vdots \\ R_{M1}\rho(t) & R_{M2}\rho(t) & R_{M3}\rho(t) & \cdots & R_{MM}\rho(t) \end{pmatrix}, \tag{8.28}$$

where R_{ii} is the probability of the reservoir being in state i, and R_{ij} is a measure of the coherence between reservoir states i and j. This form is true at $t = 0$, but at later times the reservoir and system will generally become entangled.

$\rho(t)$ is the reduced density matrix that is a representation of the statistical mixture of states that the system S is in. Following the general treatment, $\rho(t)$ takes the form

$$\rho(t) = \begin{pmatrix} \rho_{11} & \rho_{12} & \rho_{13} & \cdots & \rho_{1N} \\ \rho_{21} & \rho_{22} & \rho_{23} & \cdots & \rho_{2N} \\ \rho_{31} & \rho_{32} & \rho_{33} & \cdots & \rho_{3N} \\ \vdots & \vdots & \vdots & \vdots & \vdots \\ \rho_{N1} & \rho_{N2} & \rho_{N3} & \cdots & \rho_{NN} \end{pmatrix}, \tag{8.29}$$

where, similar to $\chi(t)$, ρ_{ii} is the probability of the system S being in state i, and ρ_{ij} is a measure of the coherence between system states i and j. From equation (8.28), we see that the reduced density matrix is really just the trace of the density matrix over the reservoir states

$$\rho(t) = Tr_R[\chi(t)] = \left[\sum_i R_{ii}\right]\rho(t) = 1\rho(t), \tag{8.30}$$

and that the trace of the reduced density matrix evaluates as

$$Tr_S[\rho(t)] = \sum_i \rho_{ii}(t) = 1. \tag{8.31}$$

To calculate expectation values for operators within the basis of S, we take the trace of the product of $\chi(t)$ and the operator $\hat{O}$:

$$\langle \hat{O} \rangle = \mathrm{Tr}_{S\otimes R}[\hat{O}\chi(t)]. \tag{8.32}$$

This reduces to

$$\begin{aligned} Tr_S[Tr_R[\hat{O}\chi(t)]] = Tr_S[\hat{O}Tr_R[\chi(t)]] &= Tr_S[\hat{O}\rho(t)] \\ &= \mathrm{Tr}[\hat{O}\rho(t)] \end{aligned} \tag{8.33}$$

because $\hat{O}$ does not act upon the reservoir basis.

Although we know that the reduced density matrix $\rho(t)$ (from here on referred to as the density matrix, as $\chi(t)$ serves us no computational purpose) describes the state of the system, we do not yet know anything about the time evolution of the density matrix, and hence know nothing about the evolution of the system. Using the definition of the density operator (equation (8.27)) and the Schrödinger equation, we see

$$\begin{aligned}\dot{\rho} &= \sum_i p_i [|\dot{\psi}_i\rangle\langle\psi_i| + |\psi_i\rangle\langle\dot{\psi}_i|] \\ &= \sum_i p_i \left[-\frac{i}{\hbar}H_S|\psi_i\rangle\langle\psi_i| + \frac{i}{\hbar}|\psi_i\rangle\langle\psi_i|H_S\right] \\ &= -\frac{i}{\hbar}[H_S, \rho],\end{aligned} \tag{8.34}$$

where H_S is the Hamiltonian for the system only, containing no terms that act on the reservoir, nor any interaction terms Thus, if the system Hamiltonian is known to us, we can solve for the time evolution of the system at hand. We still have not, however, added dissipation into our formalism; we have simply set up a framework into which dissipative terms may possibly be added in a clear, consistent manner. For the consideration of dissipation, we must generalize the form of equation (8.34).

8.3 The master equation with dissipation

An open quantum system typically consists of a number of atoms and field modes coupled with an environment with many degrees of freedom. While in principle the evolution of all the environmental degrees of freedom could be followed, it is computationally expensive, and detailed information about the environment is not of interest. We need a formalism which properly models the action of the environment on the system, but which does not require us to keep detailed information about the environment. (For example, we wish to know that an atom has emitted a photon, but not which direction the photon went.) One way to do this is to derive a master equation where the time evolution of the density operator is given by a Liouvillian operator

$$\dot{\rho} = \mathcal{L}\rho \tag{8.35}$$

describing both reversible Hamiltonian evolution and irreversible, dissipative processes. A brief outline of the derivation follows, and then we will present a more detailed derivation. The approach we follow is inspired by the works of Carmichael [2–4], Bruer and Pettrucione [1], and Gardiner and Zoller [5].

In a typical cavity QED system, we are concerned with the dynamics of one or many two-level atoms and the quantized electromagnetic field within a lossy optical cavity. Energy is being provided to this system via the driving field, a laser treated classically (in this work) coming through the cavity's input mirror. Dissipation arises in the system through the coupling of the atoms and the cavity field modes to an environment with many degrees of freedom. In general, we are not concerned with the details of this reservoir's evolution, we only care about the reservoir's effect on the atom-cavity system, for it is the source of dissipation. To formally treat the system's evolution, we use the master equation,

$$\dot{\rho} = \mathcal{L}\rho, \tag{8.36}$$

where $\mathcal{L}$ is a generalized Liouvillian superoperator (it operates on other operators, not states) [2]. The Liouvillian contains within it the same Hamiltonian evolution given by equation (8.34), but also contains the contributions of dissipative processes.

The reservoir that our system couples to is modeled as a collection of quantized harmonic oscillator modes in thermal equilibrium at some temperature. This allows us to sufficiently model many different environments. In this work, the environment is taken to be a reservoir of electromagnetic field modes. To arrive at the final Liouvillian evolution of our system, we essentially do the following:

(1) couple the system to the reservoir;
(2) use the Born and Markov approximations dealing with the system/reservoir's coupling strength, and response time of the reservoir;
(3) trace over the reservoir states to extract the system dynamics.

One type of dissipation we will examine is the decay of the cavity field mode into the vacuum radiation field outside the optical cavity, the coupling taking place via the mirror. This system is characterized by the Hamiltonian

$$H_S = \hbar\omega_0 a^\dagger a. \tag{8.37}$$

The reservoir Hamiltonian is given by

$$H_R = \sum_j \hbar\omega_j r_j^\dagger r_j, \tag{8.38}$$

where a and $a^\dagger$ (r_j and $r_j^\dagger$) are the annihilation and creation operators for the cavity field mode (jth reservoir field mode). The interaction Hamiltonian between the cavity field mode and the reservoir in the rotating wave approximation (RWA) is given by

$$H_{SR} = \sum_j \hbar\left(\kappa_j^* a r_j^\dagger + \kappa_j a^\dagger r_j\right), \tag{8.39}$$

where κ_j is the coupling strength between the cavity field mode and the jth mode of the reservoir field. The first approximation we make in solving for the system's evolution is the Born approximation. The Born approximation enters the formal derivation as a statement that the full density matrix can be factored as

$$\tilde{\chi}(t) = \tilde{\rho}(t)R_0, \tag{8.40}$$

where the tilde above $\rho(t)$ and $\chi(t)$ denote that this equation takes place in the interaction picture and not the Schrödinger picture that we have worked in thus far.

Equation (8.40) is equivalent to saying that the reservoir's properties are independent of time. In this work, this means that due to the reservoir's many degrees of freedom it is taken that the system couples only very weakly to any one mode of the vacuum field ($\kappa_j \ll 1$ for all j). Thus, the system has very little effect on the reservoir, while the reservoir modes collectively have a large effect on the system. A simple analogy might involve dropping an ice cube into the Atlantic Ocean, the ocean effects the ice cube profoundly, but the ice cube disturbs the ocean to a very small degree. The next approximation made is the Markoff approximation; a Markoffian system's future dynamics only depend on the system's present state. This approximation enters the formal derivation in the form of a qualitative comparison between the reservoir's time scale and that of the system. If the reservoir is able to return to its steady-state very quickly as compared with the time scale of the system's dynamics, then the reservoir will 'have no memory' of past interactions. This irreversibility is then in line with our intuition of dissipative processes. After these approximations are made in the formal derivation, one can then trace over the reservoir's states to obtain the Liouvillian evolution of the system's density matrix in the form of the master equation. A more formal discussion of the Born–Markoff approximation, and a derivation of the Master equation is presented in a later section.

The Liouvillian evolution of a single cavity field mode coupled to the vacuum radiation field given by equation (8.38) is

$$\begin{aligned}\dot{\rho}(t) = &-i\omega_0[a^\dagger a, \rho] + \kappa(\bar{n} + 1)[2a\rho(t)a^\dagger - \rho(t)a^\dagger a - a^\dagger a\rho(t)] \\ &+ \kappa\bar{n}[2a\rho(t)a^\dagger - aa^\dagger\rho(t) - \rho(t)aa^\dagger],\end{aligned} \tag{8.41}$$

where κ and $\bar{n}$ are the cavity field decay rate and the thermal photon number (the mean number of photons in the reservoir at thermal equilibrium), respectively.

To see that equation (10.69) gives results that are consistent with our intuition of a decaying field, we calculate the time dependent photon number of the cavity. Using equation (8.32) to calculate the time derivative of the photon number's expectation value, we find

$$\begin{aligned}\langle \dot{n} \rangle &= \mathrm{Tr}[a^\dagger a\dot{\rho}(t)] \\ &= -i\omega_0\,\mathrm{Tr}(a^\dagger a a^\dagger a\rho - a^\dagger a\rho a^\dagger a) + \kappa\,\mathrm{Tr}(2a^\dagger a^2\rho a^\dagger - a^\dagger a a^\dagger a\rho - a^\dagger a\rho a^\dagger a) \\ &\quad +2\kappa\bar{n}\,\mathrm{Tr}(a^\dagger a^2\rho a^\dagger + a^\dagger a a^\dagger\rho a - a^\dagger a a^\dagger a\rho - a^\dagger a\rho a a^\dagger) \\ &= 2\kappa\,\mathrm{Tr}((a^\dagger)^2 a^2\rho - (a^\dagger a)^2\rho) \\ &\quad + 2\kappa\bar{n}\,\mathrm{Tr}((a^\dagger)^2a^2\rho + (a^\dagger a)^2\rho - (a^\dagger a)^2\rho - a(a^\dagger)^2 a\rho) \\ &= -2\kappa[\langle n\rangle - \bar{n}].\end{aligned} \tag{8.42}$$

This differential equation has the solution

$$\langle n(t)\rangle = \langle n(0)\rangle e^{-2\kappa t} + \bar{n}(1 - e^{-2\kappa t}). \tag{8.43}$$

From this example, we see that our method provides us with a photon number that starts at some initial value $\langle n(0)\rangle$ and evolves to its thermal steady state value $\bar{n}$; thus, our method is in congruence with our intuition.

Another type of dissipation that plays an important role in the evolution of a cavity QED system is the decay of the atomic inversion, that is radiative decay of the two-level atom. Here, our system Hamiltonian is given by

$$H_S = \frac{1}{2}\hbar\omega_0\sigma_z, \tag{8.44}$$

where σ_z is the Pauli pseudo-spin operator that describes the atomic inversion

$$\sigma_z = \begin{pmatrix} 1 & 0 \\ 0 & -1 \end{pmatrix}. \tag{8.45}$$

The environment that we will couple the system to is a reservoir of cavity field modes, which has the same Hamiltonian given by equation (8.38); the interaction Hamiltonian is

$$H_{SR} = \sum_{\vec{k},p}\hbar\left(\kappa^*_{\vec{k},p} r^\dagger_{\vec{k},p}\sigma_- + \kappa_{\vec{k},p} r_{\vec{k},p}\sigma_+\right), \tag{8.46}$$

where $\vec{k}$ denotes the wavevector of the field mode and p denotes the mode's polarization. σ_+ and σ_- are the Pauli pseudo-spin operators that take a two-level atom from the excited state to the ground state and the atom from the ground state to the excited state, respectively

$$\sigma_+ = \begin{pmatrix} 0 & 1 \\ 0 & 0 \end{pmatrix}, \tag{8.47}$$

$$\sigma_- = \begin{pmatrix} 0 & 0 \\ 1 & 0 \end{pmatrix}. \tag{8.48}$$

Following a derivation analogous to that for the damped cavity field, and making the same approximations, we find that the master equation for a radiatively damped two-level atom is given by

$$\begin{aligned} \dot{\rho} = &-\frac{i}{2}\omega_0[\sigma_z, \rho] + \frac{\gamma}{2}(\bar{n}_a + 1)(2\sigma_-\rho\sigma_+ - \sigma_+\sigma_-\rho - \rho\sigma_+\sigma_-) \\ &+ \frac{\gamma}{2}\bar{n}_a(2\sigma_+\rho\sigma_- - \sigma_-\sigma_+\rho - \rho\sigma_-\sigma_+), \end{aligned} \tag{8.49}$$

where $\bar{n}_a$ is now the thermal photon number of the reservoir that the atom couples to, and γ now denotes the rate of spontaneous emission.

At the optical wavelengths often considered in cavity QED experiments, the thermal photon number $\bar{n}$ is often very small and can be ignored. At $\omega_0 = 3 \times 10^{15}$, $T = 300\ K$, $\bar{n} \approx 7 \times 10^{-34} \approx 0$. Taking this approximation, we will show the simple calculation of the spontaneous emission dynamics of the two-level atom. By the definition of the reduced density matrix $\rho(t)$, the probability of finding the two-level atom in its excited state at time t is given by the population

$$\rho_{ee}(t) = \langle e|\rho(t)|e\rangle. \tag{8.50}$$

Thus, the time derivative of this population is given by

$$\begin{aligned}\dot{\rho}_{ee} &= \langle e|\dot{\rho}_{ee}|e\rangle \\ &= -\frac{i}{2}\omega_0\langle e|(\sigma_z\rho - \rho\sigma_z)|e\rangle + \frac{\gamma}{2}\langle e|(2\sigma_-\rho\sigma_+ - \sigma_+\sigma_-\rho - \rho\sigma_+\sigma_-)|e\rangle \\ &= -\gamma\rho_{ee}.\end{aligned} \tag{8.51}$$

This is a familiar differential equation, with the familiar solution

$$\rho_{ee}(t) = \rho_{ee}(0)e^{-\gamma t}. \tag{8.52}$$

This result matches the classical result and is consistent with our intuition of radiative decay.

8.4 Quantum regression theorem

We have seen earlier that aspects of optical spectra and photon statistics are described by two-time correlation functions. For example, the power spectrum of light is given by

$$\begin{aligned}I(\omega) &= \frac{T}{2\pi}\int_{-\infty}^{\infty} d\tau e^{-i\omega\tau}\langle a^\dagger(t+\tau)a(t)\rangle \\ &= \frac{\langle a^\dagger a\rangle T}{2\pi}\int_{-\infty}^{\infty} d\tau e^{-i\omega\tau}g^{(1)}(\tau)\end{aligned} \tag{8.53}$$

Also, the probability of detecting a photon at time τ given that one was detected at $t = 0$, scaled by the probability of detecting two independent photons (randomly generated via a Poisson process, as would result from a coherent state),

$$g^{(2)}(\tau) = G^{(2)}(\tau)/G^{(2)}(\infty) \tag{8.54}$$

$$= \frac{\langle a^\dagger(0)a^\dagger(\tau)a(\tau)a(0)\rangle}{\langle a^\dagger a\rangle_{ss}^2}. \tag{8.55}$$

The key to calculating correlation functions from the master equation is the quantum regression theorem. As we have seen, correlation functions take the form

$$\langle\hat{O}_1(t)\hat{O}_2(t+\tau)\rangle = \mathrm{tr}\left\{\hat{O}_2(0)e^{\mathcal{L}\tau}[\rho(t)\hat{O}_1(0)]\right\} \tag{8.56}$$

$$\langle\hat{O}_1(t)\hat{O}_2(t+\tau)\hat{O}_3(t)\rangle = \mathrm{tr}\left\{\hat{O}_2(0)e^{\mathcal{L}\tau}[\hat{O}_3(0)\rho(t)\hat{O}_1(0)]\right\}. \tag{8.57}$$

To calculate equation (8.56) for a given time t, we define an operator

$$A(\tau) = e^{\mathcal{L}\tau}\rho(t)\hat{O}_1(0). \tag{8.58}$$

Here, we have assumed that the statistics are stationary, A depends only on the delay time τ. We are *not* saying that everything is stationary, that nothing changes. Merely that after the system has been going for a while, we can pick any time interval, beginning at any time $\tau = 0$ we wish, and the resulting averages, variances, etc, would not change. Then we can write for a complete set of operators $\vec{A}$

$$\vec{A}(\tau) = e^{\mathcal{L}\tau}\vec{A}(0) \tag{8.59}$$

$$\dot{\vec{A}} = \mathcal{L}\vec{A} \tag{8.60}$$

Notice that $\vec{A}(\tau)$ is governed by the same master equation as $\rho(t)$ although with initial conditions which depend on $\rho(t)$. All that is needed then is to solve for $\vec{A}(\tau)$ by one of the methods described at the beginning of this section, using the result in equation (8.56). The solution of equation (8.57) follows from a similar method. In words the quantum regression theorem states that two-time correlation functions obey the same equations of motion with respect to τ as expectation values obey with respect to t. Often in quantum optics we are interested in the spectrum or photon statistics of the field after the system has settled down to some steady state. In this case, we let $t \to \infty$ so that $\rho(t)$ is replaced by ρ_{ss} and then correlation functions depend only on τ.

The quantum regression theorem provides a method with which to formally represent and solve two-time correlation functions, that is, an expectation value of the product of two operators evaluated at two different times. In the Heisenberg picture, this expectation value is given by

$$\langle \hat{O}_1(t)\hat{O}_2(t')\rangle = \mathrm{Tr}_{S\oplus R}[\chi(0)\hat{O}_1(t)\hat{O}_2(t')]. \tag{8.61}$$

Defining $\tau = t' - t$, using the cyclic property of the trace, and tracing over the reservoir states eventually leads to the result

$$\langle \hat{O}_1(t)\hat{O}_2(t+\tau)\rangle = Tr_S[\hat{O}_2(0)e^{\mathcal{L}\tau}[\rho(t)\hat{O}_1(0)]]. \tag{8.62}$$

If we consider time = t to be the system steady-state, and exploit statistical stationarity, we may write equation (8.62) as

$$\langle \hat{O}_1(0)\hat{O}_2(\tau)\rangle = Tr_S[\hat{O}_2(0)e^{\mathcal{L}\tau}[\rho_{ss}\hat{O}_1(0)]]. \tag{8.63}$$

Next, we define a set of operators operator

$$\vec{A}(\tau) = e^{\mathcal{L}\tau}\rho_{ss}\hat{O}_1(0), \tag{8.64}$$

where we only see the evolution of A in τ, with initial condition

$$A(0) = \rho_{ss}\hat{O}_1(0). \tag{8.65}$$

Now we can write

$$A(\tau) = e^{\mathcal{L}\tau}A(0), \tag{8.66}$$

$$\dot{A} = \mathcal{L}A. \tag{8.67}$$

Equation (8.67) shows that $A(\tau)$ obeys the same master equation as $\rho(t)$, with $A(0)$ having initial conditions dependent upon ρ_{ss}. The familiar result of this is that the two-time correlation function in equation (8.63) satisfies the same equation of motion with respect to τ as expectation values do with respect to t.

$$\langle \hat{O}_1(0)\hat{O}_2(\tau)\rangle = Tr_S[\hat{O}_2(0)A(\tau)]. \tag{8.68}$$

8.5 Derivation of the master equation in the Born–Markoff approximation

To begin the formal derivation of the master equation in the Born–Markoff approximation, we start with a generalized form of equation (8.34) that is applicable to the full density operator, $\chi(t)$

$$\dot{\tilde{\chi}}(t) = \frac{i}{\hbar}[\tilde{\chi}(t), H_i], \tag{8.69}$$

where, as before, tilde denotes that we're working in the interaction picture and H_i is the interaction Hamiltonian for the system and the reservoir(s). We now formally integrate equation (8.69) and solve for $\tilde{\chi}(t)$

$$\tilde{\chi}(t) = \tilde{\chi}(0) + \frac{i}{\hbar}\int_0^t dt'[\tilde{\chi}(t'), H_i(t')]. \tag{8.70}$$

We now substitute this back into the right-hand side of equation (8.69) to get

$$\dot{\tilde{\chi}}(t) = \frac{i}{\hbar}[\tilde{\chi}(0), H_i] - \frac{1}{\hbar^2}\int_0^t dt'[[\tilde{\chi}(t'), H_i(t')], H_i(t)]. \tag{8.71}$$

We now make the assumption that at $t = 0$ the full density operator factorizes as

$$\tilde{\chi}(0) = \tilde{\rho}(0)R_0, \tag{8.72}$$

where $\tilde{\rho}$ is the reduced density operator in the interaction picture and R_0 is the density operator of the reservoir(s) at $t = 0$. This assumption says that at $t = 0$ we have chosen that there is no interaction between the system and the reservoir(s). We now trace over the reservoir(s) states to obtain an equation of motion for $\tilde{\rho}$ alone

$$\dot{\tilde{\rho}} = \frac{i}{\hbar}Tr_R[\tilde{\rho}(0)R_0(r), H_i(t)] - \frac{1}{\hbar^2}Tr_R\int_0^t dt'[[\tilde{\chi}(t'), H_i(t')], H_i(t)] \tag{8.73}$$

For simplicity, we assume $Tr_R[R_0(r), H_i(t)] = 0$. This just means that the mean value of the interaction Hamiltonian with respect to the reservoir is zero. This leaves us with

$$\dot{\tilde{\rho}} = -\frac{1}{\hbar^2}Tr_R\int_0^t dt'[[\tilde{\chi}(t'), H_i(t')], H_i(t)]. \tag{8.74}$$

We now state the Born approximation, that to lowest order in H_i the dynamics of the full density operator only depend upon the dynamics of the system's reduced density operator.

$$\tilde{\chi}(t) = \tilde{\chi}(t)R_0 + \mathcal{O}(H_i). \tag{8.75}$$

We may make this assumption because, as discussed in chapter 2, we take the coupling between the system and the reservoir to be small. With this weak coupling, the reservoir is taken to be unaffected by the system due to its vast size. Our system however, because of its small size as compared to the reservoir, is greatly affected by the weak coupling to many reservoir modes. Using the Born approximation, to lowest order in H_i, equation (8.74) becomes

$$\dot{\tilde{\rho}} = -\frac{1}{\hbar^2} Tr_R \int_0^t dt' [[\tilde{\rho}(t')R_0, H_i(t')], H_i(t)]. \tag{8.76}$$

It is instructive to complete the derivation with an example such that the form of the Liouvillian superoperator is justified in the equation $\dot{\rho} = \mathcal{L}\rho$. In order to do this, we must first further specify the nature of the interaction Hamiltonian H_i to be a collection of products between system and reservoir operators

$$H_i = \hbar \sum_{j,k} S_k \Gamma_{j,k}, \tag{8.77}$$

where S_k are the system operators and $\Gamma_{j,k}$ are the operators of the reservoir only. Notice here that the term we neglected in equation (8.73) is identically zero, as it would be the average of linear reservoir system operators. If these are field modes we are saying that the average field is zero. Quantum mechanics forbids us from having the variance of the field be zero. Equation (8.76) now becomes

$$\dot{\tilde{\rho}} = -\frac{1}{\hbar^2} Tr_R \int_0^t dt' \left[\left[\tilde{\rho}(t')R_0, \hbar \sum_{j,k} S_k(t')\Gamma_{j,k}(t')\right], \hbar \sum_{j,k} S_k(t)\Gamma_{j,k}(t)\right]. \tag{8.78}$$

Expanding equation (8.78) and accounting for the sum indexing, we have

$$\begin{aligned}
\dot{\tilde{\rho}} = - \sum_{j,k,l,m} \int_0^t dt' \Big[& Tr_R\big[\tilde{\rho}(t')R_0 S_k(t')\Gamma_{j,k}(t')S_m(t)\Gamma_{l,m}(t)\big] \\
& + Tr_R\big[S_k(t')\Gamma_{j,k}(t')\tilde{\rho}(t')R_0 S_m(t)\Gamma_{l,m}(t)\big] \\
& + Tr_R\big[S_m(t)\Gamma_{l,m}(t)\tilde{\rho}(t')R_0 S_k(t')\Gamma_{j,k}(t')\big] \\
& + Tr_R\big[S_m(t)\Gamma_{l,m}(t)S_k(t')\Gamma_{j,k}(t')\tilde{\rho}(t')R_0\big]\Big].
\end{aligned} \tag{8.79}$$

After performing the trace over the reservoir states, equation (8.79) simplifies to

$$\begin{aligned}\dot{\tilde{\rho}} = - \sum_{j,k,l,m} \int_0^t dt' \Big[& [S_k(t')\tilde{\rho}(t')S_m(t) - \tilde{\rho}(t')S_k(t')S_m(t)] \\ & \langle \Gamma_{j,k}(t')\Gamma_{l,m}(t)\rangle_R \\ & + [S_m(t)\tilde{\rho}(t')S_k(t') - S_m(t)S_k(t')\tilde{\rho}(t')]\langle \Gamma_{l,m}(t)\Gamma_{j,k}(t')\rangle_R \Big]. \end{aligned} \tag{8.80}$$

We are now in a position to specify the system and reservoir operators in equation (8.77) so as to work the example of a single cavity field mode coupled to a reservoir of harmonic oscillator modes. We define the creation and annihilation operators for the cavity field mode in the interaction picture

$$\begin{aligned}\tilde{S}_1(t) &= e^{i\omega_0 a^\dagger a t} a^\dagger e^{-i\omega_0 a^\dagger a t} = a^\dagger e^{-i\omega_0 t}, \\ \tilde{S}_2(t) &= e^{i\omega_0 a^\dagger a t} a e^{-i\omega_0 a^\dagger a t} = a e^{+i\omega_0 t}.\end{aligned} \tag{8.81}$$

And we define the reservoir oscillator mode operators to be

$$\begin{aligned}\tilde{\Gamma}_{1,j}(t) &= e^{i\omega_j c_j^\dagger c_j t} \kappa_j c_j e^{-i\omega_j c_j^\dagger c t} = \kappa_j c_j e^{-i\omega_j t}, \\ \tilde{\Gamma}_{2,j}(t) &= e^{i\omega_j c_j^\dagger c_j t} \kappa_j^* c_j^\dagger e^{-i\omega_j c_j^\dagger c t} = \kappa_j^* c_j^\dagger e^{+i\omega_j t}.\end{aligned} \tag{8.82}$$

Using these operator definitions, we may now bring the sum through equation (8.80) to arrive at

$$\begin{aligned}\dot{\tilde{\rho}} = \int_0^t dt' \Bigg(& [a\tilde{\rho}(t')a - aa\tilde{\rho}(t')]e^{-i\omega_0(t+t')} \sum_{j,l} \langle \Gamma_{2,j}(t')\Gamma_{2,l}(t)\rangle_R + h.c. \\ & + [a^\dagger\tilde{\rho}(t')a^\dagger - a^\dagger a^\dagger\tilde{\rho}(t')]e^{+i\omega_0(t+t')} \sum_{j,l} \langle \Gamma_{1,j}(t')\Gamma_{1,l}(t)\rangle_R + h.c. \\ & + [a^\dagger\tilde{\rho}(t')a - aa^\dagger\tilde{\rho}(t')]e^{-i\omega_0(t-t')} \sum_{j,l} \langle \Gamma_{2,j}(t)\Gamma_{1,l}(t')\rangle_R + h.c. \\ & + [a\tilde{\rho}(t')a^\dagger - a^\dagger a\tilde{\rho}(t')]e^{-i\omega_0(t+t')} \sum_{j,l} \langle \Gamma_{1,j}(t)\Gamma_{2,l}(t')\rangle_R + h.c. \Bigg).\end{aligned} \tag{8.83}$$

where $h.c.$ denotes the Hermitian conjugate of the term immediately preceding. The specific forms of the reservoir correlations needed to evaluate equation (8.83) are

$$\begin{aligned}
\sum_{j,l}\langle\tilde{\Gamma}_{2,j}(t')\tilde{\Gamma}_{2,l}(t)\rangle &= \sum_{j,l}\kappa_j^*\kappa_l^* e^{i\omega_j t'} e^{i\omega_l t} Tr_R(R_0 c_j^\dagger c_l^\dagger) \\
&= 0, \\
\sum_{j,l}\langle\tilde{\Gamma}_{1,j}(t)\tilde{\Gamma}_{1,l}(t')\rangle &= \sum_{j,l}\kappa_j\kappa_l e^{-i\omega_j t'} e^{-i\omega_l t} Tr_R(R_0 c_j c_l) \\
&= 0, \\
\sum_{j,l}\langle\tilde{\Gamma}_{2,j}(t)\tilde{\Gamma}_{1,l}(t')\rangle &= \sum_{j,l}\kappa_j^*\kappa_l e^{+i\omega_j t} e^{-i\omega_l t'} Tr_R(R_0 c_j^\dagger c_l) \\
&= \sum_j |\kappa_j|^2 e^{+i\omega_j(t-t')}\bar{n}(\omega_j, T), \\
\sum_{j,l}\langle\tilde{\Gamma}_{1,j}(t)\tilde{\Gamma}_{2,l}(t')\rangle &= \sum_{j,l}\kappa_j\kappa_l^* e^{-i\omega_j t} e^{+i\omega_l t'} Tr_R(R_0 c_j c_l^\dagger) \\
&= \sum_j |\kappa_j|^2 e^{-i\omega_j(t-t')}\left(\bar{n}(\omega_j, T) + 1\right),
\end{aligned} \tag{8.84}$$

where $\bar{n}(\omega_j, T)$ is the average thermal photon number for reservoir mode j in thermal equilibrium at temperature T as discussed in chapter 2. More specifically,

$$\begin{aligned}
\bar{n}(\omega_j, T) &= Tr_R\left(R_0 c_j^\dagger c_j\right) \\
&= \langle c_j^\dagger c_j\rangle \\
&= \frac{e^{-\hbar\omega_j/k_B T}}{1 - e^{-\hbar\omega_j/k_B T}}.
\end{aligned} \tag{8.85}$$

To continue, it is helpful if we make a change of variables and an approximation about the nature of the reservoir's mode spacing. First we define the delay time $\tau = t - t'$. Next, we state that the reservoir's modes are very tightly spaced, warranting the conversion of the sum in equations (8.84) to integrals. Defining the *mode density* $g(\omega)$ such that $g(\omega)d\omega$ is the number of reservoir modes of frequencies between ω and $\omega + d\omega$ and using τ in place of $t - t'$, we have

$$\begin{aligned}
\sum_j |\kappa_j|^2 e^{+i\omega_j(t-t')}\bar{n}(\omega_j, T) &\Rightarrow \int_0^\infty d\omega e^{+i\omega\tau} g(\omega)|\kappa(\omega)|^2 \bar{n}(\omega, T), \\
\sum_j |\kappa_j|^2 e^{-i\omega_j(t-t')}(\bar{n}(\omega_j, T) + 1) &\Rightarrow \int_0^\infty d\omega e^{-i\omega\tau} g(\omega)|\kappa(\omega)|^2 (\bar{n}(\omega, T) + 1).
\end{aligned} \tag{8.86}$$

Equation (8.83) may now be written as

$$\begin{aligned}\dot{\tilde{\rho}} = \int_0^t d\tau\bigg(&[a^\dagger\tilde{\rho}(t-\tau)a - aa^\dagger\tilde{\rho}(t-\tau)] \\ &\int_0^\infty d\omega e^{+i(\omega-\omega_0)\tau} g(\omega)|\kappa(\omega)|^2 \bar{n}(\omega, T) + \text{h.c.} \\ &+ [a\tilde{\rho}(t-\tau)a^\dagger - a^\dagger a\tilde{\rho}(t-\tau)] \\ &\int_0^\infty d\omega e^{-i(\omega-\omega_0)\tau} g(\omega)|\kappa(\omega)|^2 (\bar{n}(\omega, T) + 1) + \text{h.c.}\bigg).\end{aligned} \tag{8.87}$$

Now, we are able to make the formal statement of the Markoff approximation. The integrals of equations (8.86) are essentially delta functions, as any sizable τ will average the 'slower varying' functions $g(\omega)$, $|\kappa(\omega)|^2$ and $\bar{n}(\omega, T)$ to zero. This means that the reservoir correlations 'die off' on a time scale much shorter than the natural time scale of the cavity field mode system (and hence $\tilde{\rho}$). We may therefore replace $\tilde{\rho}(t-\tau)$ with $\tilde{\rho}(t)$ and extend the upper limit on the time integral of equation (8.87) to infinity. We may now write

$$\lim_{t\to\infty} \int_0^t d\tau e^{-i(\omega-\omega_0)\tau} \approx \pi\delta(\omega - \omega_0), \tag{8.88}$$

where we have ignored a small, imaginary offset (this offset is equivalent to the Lamb shift for our cavity field mode 'oscillator'). This delta function reduces the remaining integral over ω, and we may now write equation (8.87) as

$$\begin{aligned}\dot{\tilde{\rho}} = &\frac{\kappa'}{2}(2a\tilde{\rho}a^\dagger - a^\dagger a\tilde{\rho} - \tilde{\rho}a^\dagger a) \\ &+ \kappa'\bar{n}(a\tilde{\rho}a^\dagger + a^\dagger\tilde{\rho}a - a^\dagger a\tilde{\rho} - \tilde{\rho}aa^\dagger),\end{aligned} \tag{8.89}$$

where

$$\begin{aligned}\kappa' &= 2\pi g(\omega_0)|\kappa(\omega_0)|^2, \\ \bar{n} &\equiv \bar{n}(\omega_0, T).\end{aligned} \tag{8.90}$$

We may transform this equation back into the Schrödinger picture by the usual prescription,

$$\dot{\rho} = \frac{-i}{\hbar}[H_S, \rho] + e^{-(i/\hbar)H_S t}\dot{\tilde{\rho}}e^{+(i/\hbar)H_S t}, \tag{8.91}$$

to obtain the familiar result, as presented in chapter 2 (using the Linblad form)

$$\begin{aligned}\dot{\rho} = &-i\omega_0[a^\dagger a, \rho] + \frac{\kappa'}{2}(\bar{n}+1)(2a\rho a^\dagger - a^\dagger a\rho - \rho a^\dagger a) \\ &+ \frac{\kappa'}{2}\bar{n}(2a^\dagger\rho a - aa^\dagger\rho - \rho aa^\dagger).\end{aligned} \tag{8.92}$$

8.6 Other types of reservoirs

We can also have the atomic inversion coupled to a heat bath. Here, the interaction is given by

$$H_{SB} = \sum_k (a_k + a_k^\dagger)\sigma_z \tag{8.93}$$

This results in the master equation addition

$$\dot{\rho} = -\Gamma(\sigma_z\rho\sigma_z - \rho) \tag{8.94}$$

This does not seem in Lindblad form, but it can be written as

$$\dot{\rho} = -\frac{\Gamma}{2}(\sigma_z\rho\sigma_z - \sigma_z^2\rho - \rho\sigma_z^2) \tag{8.95}$$

which is identical, once we realize that $\sigma_z^2 = 1$.

Also, we can have phase dependent reservoirs where

$$\langle a_k a_k \rangle \tag{8.96}$$

is not zero. For a bath of squeezed vacuum states, we have

$$Tr_R(Rc_j^\dagger c_j) = N \tag{8.97}$$

$$Tr_R(Rc_j c_j^\dagger) = N + 1 \tag{8.98}$$

$$Tr_R(Rc_j c_j) = M \tag{8.99}$$

$$Tr_R(Rc_j^\dagger c_j^\dagger) = M^* \tag{8.100}$$

where we must have

$$|M|^2 \leqslant N(N + 1) \tag{8.101}$$

For a squeezed vacuum, we have $N = \sinh^2(r)$, and $M = \sinh(r)\cosh(r)e^{\phi}$, where ϕ is the phase of the squeezed vacuum, related to the direction the fluctuations are reduced. This produces a master equation of the form

$$\begin{aligned}\dot{\rho}(t) = &-i\omega_0[a^\dagger a, \rho] + \kappa(N + 1)[2a\rho(t)a^\dagger - \rho(t)a^\dagger a - a^\dagger a\rho(t)]\\ &+ \kappa N[2a\rho(t)a^\dagger - aa^\dagger\rho(t) - \rho(t)aa^\dagger]\end{aligned} \tag{8.102}$$

$$+ \kappa M[2a\rho(t)a - a^2\rho(t) - \rho(t)a^2] \tag{8.103}$$

$$+ \kappa M^*[2a^\dagger\rho(t)a^\dagger - a^{\dagger 2}\rho(t) - \rho(t)a^{\dagger 2}] \tag{8.104}$$

The corresponding equation for atomic decay into a squeezed reservoir is simpler as $\sigma_\pm^2 = 0$

$$\dot{\rho} = -\frac{i}{2}\omega_0[\sigma_z, \rho] + \frac{\gamma}{2}(N+1)(2\sigma_-\rho\sigma_+ - \sigma_+\sigma_-\rho - \rho\sigma_+\sigma_-) \\ + \frac{\gamma}{2}N_a(2\sigma_+\rho\sigma_- - \sigma_-\sigma_+\rho - \rho\sigma_-\sigma_+) \tag{8.105}$$

$$+ \frac{\gamma}{2}M(2\sigma_\rho\sigma_-) \tag{8.106}$$

$$+ \frac{\gamma}{2}M^*(2\sigma_+\rho\sigma_+) \tag{8.107}$$

8.7 Alternative derivation of Lindblad equation

The appropriate generalization of the Schrödinger equation was worked out in the early 1960s [6, 7]. To see how it arises, we follow a well-trodden path, consider the combined state of a system of interest and a reservoir that the system interacts with described by a density operator ρ and assume that initially the system and the reservoir are not entangled so we may write

$$\rho = \rho_s \otimes R, \tag{8.108}$$

where ρ_s depends only on the state of the system and

$$R = \sum_n P_n|n\rangle\langle n|, \tag{8.109}$$

describes the state of the reservoir. In order to find a description that only references the state of the system we trace over the reservoir degrees of freedom. The system density operator at a time t is then related to the initial system density operator by

$$\rho_s(t) = \mathrm{Tr}_R(U\rho_s \otimes P_n|n\rangle\langle n|U^\dagger) = \sum_\mu \sum_n \sqrt{P_n}\langle\mu|U|n\rangle\rho_s(0)\langle n|U^\dagger|\mu\rangle\sqrt{P_n}, \tag{8.110}$$

where the $|\mu\rangle$ span the reservoir section of the Hilbert space. Defining the Kraus operators to be

$$M_{\mu,n}(t) = M_\nu(t) = \langle\mu|U|n\rangle, \tag{8.111}$$

where the first equal sign is a relabeling. The system density operator at time t is then given by

$$\rho_s(t) = \sum_\nu M_\nu(t)\rho_s M_\nu^\dagger(t) \equiv a(t)[\rho_s(0)] \tag{8.112}$$

This map is referred to as completely positive since it will always map a positive operator to a positive operator even if there are other subsystems present on which it must act as the identity.

An important property of this map is that it is trace preserving.

$$\begin{aligned}\mathrm{Tr}_s\big(\rho_s(t)\big) &= \mathrm{Tr}_s\left(\sum_{\mu,n} M_{\mu,n}\rho_s(0)M_{\mu,n}^{\dagger}\right)\\ &= \mathrm{Tr}_s\left(\sum_{n} P_n\langle n|U^{\dagger}\sum_{\mu}|\mu\rangle\langle\mu|U|n\rangle\rho_s(0)\right)\\ &= \mathrm{Tr}_s\big(\rho_s(0)\big).\end{aligned} \tag{8.113}$$

It is important to note that this property held true because

$$\sum_{\nu} M_{\nu}^{\dagger}M_{\nu} = \mathbf{1}_s. \tag{8.114}$$

This is sometimes referred to as the completeness property of the Kraus operators. One subtlety here is that the map is only trace preserving if the $|n\rangle$ and the $|\mu\rangle$ are capable of forming resolutions of the identity in some section of the Hilbert space.

To first order in dt, we may write one of the Kraus operators as

$$M_0 = \mathbf{1}_s + G(t)dt \tag{8.115}$$

and define a new set of operators $L_{\nu}(t)$ by

$$M_{\nu}(t) = L_{\nu}(t)\sqrt{dt}. \tag{8.116}$$

The completeness property of the Kraus operators implies

$$M_0^{\dagger}M_0 = \mathbf{1}_s - \sum_{\nu=1} M_{\nu}^{\dagger}M_{\nu}. \tag{8.117}$$

To first order in dt, this implies

$$\mathbf{1}_s + (G(t) + G^{\dagger}(t))dt = \mathbf{1}_s - \sum_{\nu=1} L_{\nu}^{\dagger}(t)L_{\nu}(t)dt \tag{8.118}$$

This implies that

$$G(t) = -\frac{1}{2}\sum_{\nu} L_{\nu}^{\dagger}(t)L_{\nu}(t) + A(t), \tag{8.119}$$

where $A(t)$ is anti-Hermitian. Plugging this into the map above and differentiating with respect to t gives

$$\dot{\rho}_s = [A, \rho_s] + \sum_{i}(2L_i\rho L_i^{\dagger} - L_i^{\dagger}L_i\rho - \rho L_i^{\dagger}L_i). \tag{8.120}$$

This suggests that a reasonable guess for $A(t)$ is

$$A(t) = -\frac{i}{\hbar}H. \tag{8.121}$$

This turns out to be correct as will be shown below. The differential equation for ρ_s then becomes

$$\dot{\rho}_s = -\frac{i}{\hbar}[H, \rho_s] + \sum_i (2L_i\rho L_i^\dagger - L_i^\dagger L_i\rho - \rho L_i^\dagger L_i), \tag{8.122}$$

where the L_i are called Lindblad or jump operators. This equation is usually referred to as the master equation in diagonal Lindblad form or simply the master equation. Sometimes it is also called the Lindbladian or the Gorini–Kossakowski–Sudarshan–Lindblad (GKSL) equation after a group of scientists who developed this formalism based on semigroups. In atomic and optical physics applications the L_i are usually operators of the form $\sqrt{\frac{\gamma}{2}}\,|i\rangle\langle j|$ where the γ are decay rates and $|j\rangle$ is a state of higher energy than $|i\rangle$. Strictly speaking however, this is only true at zero temperature because at nonzero temperatures thermal photons may cause jumps to higher energy levels. For example, the master equation that describes the evolution of two-level atom interacting with the reservoir of electromagnetic field modes is

$$\begin{aligned} \dot{\rho} = &-i\frac{\omega}{2}[\sigma_z, \rho] + \frac{\gamma}{2}(\bar{n} + 1)(2\sigma_-\rho\sigma_+ - \sigma_+\sigma_-\rho - \rho\sigma_+\sigma_-) \\ &+ \frac{\gamma}{2}\bar{n}(2\sigma_+\rho\sigma_- - \sigma_-\sigma_+\rho - \rho\sigma_-\sigma_+), \end{aligned} \tag{8.123}$$

where ω is the light shifted frequency difference between the ground and excited states, γ is the atomic decay rate, $\bar{n}$ is the thermal photon number and σ_z and $\sigma_\pm$ are the inversion and atomic raising/lowering operators, respectively. In the second term, the lowering operators play the role of the jump operators representing spontaneous and stimulated emission. In the third term the raising operators play the role of the jump operators representing absorption. Since this is an important example in atomic physics we should note that at optical frequencies and room temperature $\bar{n} \approx 0$, so we can make the zero temperature approximation and the master equation reduces to

$$\dot{\rho} = -i\frac{\omega}{2}[\sigma_z, \rho] + \frac{\gamma}{2}(2\sigma_-\rho\sigma_+ - \sigma_+\sigma_-\rho - \rho\sigma_+\sigma_-). \tag{8.124}$$

The approach most commonly taken to derive the form of $A(t)$ and the jump operators L_i, and therefore to obtain the master equation from first principles, is to write the Schrödinger equation as an integro-differential equation and then trace over the reservoir. Here, we will, instead, explicitly write down the Kraus operators using a particular basis to perform the trace, plug these into the map and then differentiate. To proceed, we must specify the state of the reservoir. By far the most common choice is for the reservoir to be a collection of harmonic oscillators in a thermal state,

$$R = \bigotimes_{\mathbf{k}} \sum_n \frac{e^{\beta n\hbar\omega_{\mathbf{k}}}}{Z}|n(\omega_{\mathbf{k}})\rangle\langle n(\omega_{\mathbf{k}})|, \tag{8.125}$$

where $\omega_{\mathbf{k}}$ is the frequency of oscillator labeled by $\mathbf{k}$, the $|n(\omega_{\mathbf{k}})\rangle$ are eigenstates of the harmonic oscillator Hamiltonian, $\beta = k_B T$, where k_B is Boltzmann's constant and T is the temperature, and Z is the partition function. We will also use these number states as a basis for the partial trace. Working in the interaction picture and to second order in dt the Kraus operators become

$$\begin{aligned} M_{\mu,n} = & \frac{e^{-\beta n \hbar \omega/2}}{\sqrt{Z}} \langle [\mu(\omega_{\mathbf{k}})] | \mathbf{1}_s + \left(-\frac{i}{\hbar}\right) \int dt H_{int}(t) \\ & + \left(-\frac{i}{\hbar}\right)^2 \int \int dt dt' H_{int}(t) H_{int}(t') | [n(\omega_{\mathbf{k}}')] \rangle, \end{aligned} \tag{8.126}$$

where the brackets, $[n(\omega_{\mathbf{k}})]$, indicate that the occupation numbers of all oscillators must be specified. The interaction Hamiltonian will be taken to be of the form

$$H_{int} = \hbar \sum_{\mathbf{k},i,j,i\neq j} g_{ij}(\omega_{\mathbf{k}}) s_{ij}^{\dagger} a_{\mathbf{k}} + g_{ij}^{*}(\omega_{\mathbf{k}}) s_{ij} a_{\mathbf{k}}^{\dagger} \tag{8.127}$$

Since each term of the interaction Hamiltonian only involves one oscillator, $\mathbf{k}$, it is convenient to split the Kraus operators up so that each one is associated with only one oscillator. There are three possible forms for these operators. First, if $|n(\omega_{\mathbf{k}})\rangle = |\mu(\omega_{\mathbf{k}})\rangle$, then

$$\begin{aligned} M_{0,n,\omega_{\mathbf{k}}} = & \frac{e^{-\beta n \hbar \omega_{\mathbf{k}}/2}}{\sqrt{Z}} \Bigg[\mathbf{1}_s - \int \int dt dt' |g(\omega_{\mathbf{k}})|^2 \\ & \sum_{i,j,i\neq j} \Big((n(\omega_{\mathbf{k}}) + 1) s_{ij}^{\dagger}(\omega, t) s_{ij}(\omega, t) + n(\omega_{\mathbf{k}}) s_{ij}(\omega, t) s_{ij}^{\dagger}(\omega, t) \Big) \Bigg], \end{aligned} \tag{8.128}$$

where the $s_{ij}(\omega, t)$ are system operators that connect state i to state j. If $|n(\omega_{\mathbf{k}})\rangle = |\mu + 1(\omega_{\mathbf{k}})\rangle$, then

$$M_{+,n,\omega} = -i \frac{e^{-\beta n \hbar \omega_{\mathbf{k}}/2}}{\sqrt{Z}} g^{*}(\omega_{\mathbf{k}}) \sqrt{n(\omega_{\mathbf{k}}) + 1} \int dt \sum_{i,j,i\neq j} s_{ij}(\omega, t). \tag{8.129}$$

Finally, if $|n, \omega'\rangle = |\mu - 1, \omega\rangle$

$$M_{-,n,\omega_{\mathbf{k}}} = -i \frac{e^{-\beta n \hbar \omega_{\mathbf{k}}/2}}{\sqrt{Z}} g(\omega_{\mathbf{k}}) \sqrt{n(\omega_{\mathbf{k}})} \int dt \sum_{i,j,i\neq j} s_{ij}^{\dagger}(\omega, t). \tag{8.130}$$

Technically, terms that involve $s_{ij}(\omega, t) s_{ij}(\omega, t)$ are possible but they are already second order in dt and we wish to eventually deal with products of Kraus operators so these will not remain second order. Accordingly, they have been dropped.

At this point, we could use these Kraus operators to write down jump operators and $A(t)$ but it is more enlightening to plug them back into the map and study the master equation as a whole. When plugging these back into the map we keep terms only to second order in dt. We must now add up all the terms associated with various values of $n(\omega_\mathbf{k})$ and $\mathbf{k}$. Two important things happen. First, we will get sums of the form $\frac{e^{-\beta n\hbar\omega}}{Z}(n+1) = \bar{n}+1$. The effect is to replace all of the n with $\bar{n}$. There will also be time integrals of the form

$$\int_t^\tau dt' e^{-i(\omega-\omega_{ij})(t'-t)}, \tag{8.131}$$

where ω_{ij} is the frequency difference between system states since in the interaction picture the operators are of the form $s_{ij}e^{-i\omega_{ij}t}$ where s_{ij} are the Schrödinger picture operators. Integrals of this form may be familiar since they emerge in the Wigner–Wiesskopf theory of spontaneous emission [8]. This integral can be evaluated by making the Markoff approximation. That is, if reservoir correlations $\langle a^\dagger a e^{-i(\omega-\omega_0)(t'-t)}\rangle$ are sharply peaked around $t = t'$ then we may extend the integral to infinity and write the integral as

$$\lim_{\tau\to\infty}\int_t^\tau dt' e^{-i(\omega-\omega_{ij})(t'-t)} = \pi\delta(\omega-\omega_{ij}) - i\mathbf{P}\left(\frac{1}{\omega-\omega_{ij}}\right), \tag{8.132}$$

where $\mathbf{P}$ denotes Cauchy's principle part. The next step is to replace the sum over $\mathbf{k}$ with an integral over ω

$$\sum_\mathbf{k} \to \int d\omega D(\omega), \tag{8.133}$$

where $D(\omega)$ is the density of states. After this coarse graining is preformed, the remaining integrals are of the form.

$$\int d\omega |g(\omega)|^2 D(\omega)(\bar{n}+1)\int_t^\infty dt' e^{-i(\omega-\omega_{ij})(t'-t)} = (\bar{n}+1)\frac{\Gamma_{ij}}{2} - \frac{i}{\hbar}\delta E_{+,ij}, \tag{8.134}$$

where

$$\Gamma_{ij} = 2\pi|g(\omega_{ij})|^2 D(\omega_{ij}), \tag{8.135}$$

and

$$\delta E_{+,ij} = \mathbf{P}\int \frac{D(\omega)(\bar{n}(\omega)+1)|g(\omega)|^2}{\omega-\omega_{ij}} d\omega. \tag{8.136}$$

Similar integrals appear for the $\bar{n}$ terms. In atomic physics, the Γ_{ij} usually are decay rates and the $\delta E_{+,ij}$ are light shifts such as the partial Lamb shift. The light shifts group with the system Hamiltonian to provide terms of the form

$$\begin{aligned}\delta H = \hbar\sum_{i,j,i\neq j} s_{ij}s_{ij}^\dagger \mathbf{P}\int \frac{D(\omega)\bar{n}(\omega)|g(\omega)|^2}{\omega-\omega_{ij}} d\omega \\ + s_{ij}^\dagger s_{ij}\mathbf{P}\int \frac{D(\omega)(\bar{n}(\omega)+1)|g(\omega)|^2}{\omega-\omega_{ij}} d\omega\end{aligned} \tag{8.137}$$

Plugging all of this back into the map and differentiating as above gives

$$\begin{aligned}\dot{\rho}_s = &-\frac{i}{\hbar}[\delta H, \rho_s] + \sum_{i,j,i\neq j} \frac{\Gamma_{ij}}{2}(\bar{n}+1)(2s_{ij}\rho s_{ij}^{\dagger} - s_{ij}^{\dagger}s_{ij}\rho - \rho s_{ij}^{\dagger}s_{ij}) \\ &+ \frac{\Gamma_{ij}}{2}\bar{n}(2s_{ij}^{\dagger}\rho s_{ij} - s_{ij}s_{ij}^{\dagger}\rho - \rho s_{ij}s_{ij}^{\dagger}).\end{aligned} \tag{8.138}$$

This is the expected form of the master equation in the interaction picture. Note that we could have explicitly written down the form of the $L_\mu(t)$ and $A(t)$ but it was not necessary. It is somewhat more common to work with the master equation in the Schrödinger picture

$$\begin{aligned}\dot{\rho}_s = &-\frac{i}{\hbar}[H_s + \delta H, \rho_s] + \sum_{i,j,i\neq j} \frac{\Gamma_{ij}}{2}(\bar{n}+1)(2s_{ij}\rho s_{ij}^{\dagger} - s_{ij}^{\dagger}s_{ij}\rho - \rho s_{ij}^{\dagger}s_{ij}) \\ &+ \frac{\Gamma_{ij}}{2}\bar{n}(2s_{ij}^{\dagger}\rho s_{ij} - s_{ij}s_{ij}^{\dagger}\rho - \rho s_{ij}s_{ij}^{\dagger}).\end{aligned} \tag{8.139}$$

As mentioned above, even in the time evolution of pure states there is one aspect that is not unitary and destroys coherent superpositions. This is measurement. This suggests that the interaction of a system with its environment can be regarded as measuring the system but not telling us about the result. This turns out to be a very useful way of thinking about this interaction.

It leads to a way of studying open quantum system time evolution called quantum trajectory theory that was independently developed in the late 1980s and early 1990s by Howard Carmichael [2–4]. If a system is continuously measured, it will remain in a pure state. An alternative method of analyzing open quantum system time evolution is then to allow the system to be represented by a ket and undergo time evolution according to the Schrödinger with a non-Hermitian Hamiltonian punctuated by quantum jumps. In atomic physics, the jumps usually correspond to the atom incoherently transitioning to a lower energy state and emitting a photon. The Hamiltonian must be non-Hermitian in order to account for the possibility that the absence of a jump indicates that the atom was already in the ground state. This perspective is usually referred to as quantum trajectory theory.

References

[1] Breuer H-P and Petruccione F 2002 *The Theory of Open Quantum Systems* (Oxford: Oxford University Press)

[2] Carmichael H J 1993 *An Open Systems Approach to Quantum Optics* (Berlin: Springer)

[3] Carmichael H J 2007 *Statistical Methods in Quantum Optics. 2: Non-Classical Fields* Theoretical and Mathematical Physics *(Berlin: Springer)*

[4] Carmichael H J 1999 *Statistical Methods in Quantum Optics. 1: Master equations and Fokker-Planck equations* Physics and Astronomy Online Library (Berlin: Springer)
[5] Gardiner C and Zoller P 2004 *Quantum Noise* (Berlin: Springer)
[6] Senitzky I 1960 Dissipation in quantum mechanics. The harmonic oscillator i *Phys. Rev.* **119** 670
[7] Senitzky I 1961 Dissipation in quantum mechanics. The harmonic oscillator ii *Phys. Rev.* **124** 642
[8] Weisskopf V and Wigner E 1930 Berechnung der naturlichen linienbreite auf grund der diracschen lichttheorie *Z. Phys.* **63** 54

Chapter 9

Quantum trajectory theory

We have seen that to formally treat the open quantum system's evolution, we can use the master equation,

$$\dot{\rho} = \mathcal{L}\rho, \tag{9.1}$$

where $\mathcal{L}$ is a generalized Liouvillian superoperator (it operates on other operators, not states) [3]. We can use it directly to calculate expectation values, and with the quantum regression theorem we can calculate two-time averages, or correlation functions. The diagonal elements of the density matrix tell us about the populations of the various states of the system, and the appearance of nonzero off-diagonal density matrix elements tells us that there is coherence of some type or another. For given initial states, we get perfectly correct answers for what is to appear on the meters in our lab; quantum mechanics actually tells us that we cannot determine *how* a system gets from initial to final state. Recently, there has been a new method developed by several workers that results in describing open systems by wave functions with back action on the system based on measurements made by detectors (real or imagined). For example one might picture an atom sitting in its excited state, and at some point a detector goes click, a photon has been detected, and then we know the atom is in its ground state. For an individual atom, one can only describe the likelihood that an atom will decay at some time. This so-called quantum jump is quite mysterious, and the ensemble nature of the density matrix theory actually hides it at first glance. In this chapter, we learn about a method that is *derived* from the general theory of open systems (not just thrown together ad hoc) that allows us to say *what we can* about process. This is akin to Feynman's sum-over-histories approach.

This other method by which we extract information from the master equation is the quantum trajectories formalism (QTF) of Carmichael [1, 2]. In the QTF, we abandon the density matrix, mixed-state representation of the system in favor of an ensemble of systems represented by pure state wavefunctions. Each of these pure

doi:10.1088/978-0-7503-1713-9ch9

state systems will undergo continuous, non-Hermitian Hamiltonian evolution that is interrupted at random times by some form of discontinuous state change. These state changes are the realizations of system decay processes, including spontaneous emission and cavity decay. The pure state representation for each member of the ensemble will take the form governed by its history of discontinuous state changes and is thus called the system's *record.* This can always be done for a master equation in Lindblad form, and is particularly useful if we wish to examine the statistics of photodetection. For other types of measurements, notably those including homodyne or heterodyne measurements, another approach will be discussed later in this chapter.

We begin with the master equation and unravel it in a particular way based on a simulated photon counting experiment. Here, we follow the formalism of Carmichael [1, 2] although there are other equivalent formulations [6]. There are also different unravelings of the master equation. One is quantum state diffusion which is based on balanced homodyne detection [4]. Nice reviews are by Wiseman [8], Jacobs and Steck [5], and Dorsselaer and Nienhuis [7].

The master equation that results in most if not all of quantum optics, where one has treated the interaction with the bath/environment by the Born–Markov approximation, is the so-called Lindblad form

$$\dot{\rho} = -i\left[H_{\text{sys}}, \rho\right] + \Gamma(2C\rho C^{\dagger} - C^{\dagger}C\rho - \rho C^{\dagger}C) \tag{9.2}$$

where C is some linear combination of creation and annihilation operators for the source. The part of master equation $-\Gamma(C^{\dagger}C\rho + \rho C^{\dagger}C)$ is an anti-commutator. By considering the evolution of a quantum state with a non-Hermitian Hamiltonian, we can accommodate these terms into a quasi-Hamiltonian evolution.

$$\dot{\rho} = |\dot{\psi}\rangle\langle\psi| + |\psi\rangle\langle\dot{\psi}| \tag{9.3}$$

$$= -\frac{i}{\hbar}H|\psi\rangle\langle\psi| + \frac{i}{\hbar}|\psi\rangle\langle\psi|H^{\dagger} \tag{9.4}$$

$$= -\frac{i}{\hbar}(H_S + i\hbar H_D)|\psi\rangle\langle\psi| + \frac{i}{\hbar}|\psi\rangle\langle\psi|(H_S - i\hbar H_D) \tag{9.5}$$

$$= -\frac{i}{\hbar}[H_S, \rho] + H_D\rho + \rho H_D \tag{9.6}$$

$$= -\frac{i}{\hbar}[H_S, \rho] + [H_D, \rho]_{+} \tag{9.7}$$

with $H_D = \hbar\Gamma C^{\dagger}C$. We can put all this together in one non-Hermitian Hamiltonian

$$H_{\text{eff}} = H_S - i\hbar\Gamma C^{\dagger}C \tag{9.8}$$

where we have

$$|\dot{\psi}\rangle = (-i/\hbar)H_{\text{eff}}|\psi\rangle \tag{9.9}$$

$$\langle\dot{\psi}| = (i/\hbar)\langle\psi|H_{\text{eff}}^{\dagger} \tag{9.10}$$

for evolution via this non-Hermitian Hamiltonian. Note that this evolution does not preserve the norm of the state vector, a point that we shall return to.

Now, the master equation is pulled apart into two pieces

$$\dot{\rho} = (\mathcal{L} - \mathcal{S})\rho + \mathcal{S}\rho \tag{9.11}$$

where we (for now) choose $\mathcal{S}$ to contain all the terms which are of the form $\hat{C}\rho\hat{C}^{\dagger}$. Each operator $\hat{C}$ in $\mathcal{S}$ becomes a collapse operator. The remaining terms are written

$$(\mathcal{L} - \mathcal{S})\rho = -\frac{i}{\hbar}[H_S, \rho] + [H_D, \rho]_+ \tag{9.12}$$

where $[\hat{O}, \rho]_+$ denotes the anticommutator and H_D contains the dissipation terms which are not included in $\mathcal{S}$. The non-Hermitian Hamiltonian which governs the trajectory evolution is

$$H = H_S + i\hbar H_D. \tag{9.13}$$

We can reproduce anticommutator terms in a density matrix evolution if we consider an underlying quantum state evolution via a non-Hermitian Hamiltonian.

How do we handle the evolution from the $\mathcal{S}$ part of the master equation? Again continuing with the idea that there is some type of wave function type object underlying ρ, we have

$$\begin{aligned}\mathcal{S}\rho &= 2\Gamma C\rho C^{\dagger}\\ &= \sqrt{2\Gamma}\, C|\psi\rangle\langle\psi|C^{\dagger}\sqrt{2\Gamma}\end{aligned} \tag{9.14}$$

where we can see the emergence of a collapse operator, $\sqrt{2\Gamma}\, C$. If, for example, this is the annihilation operator for the field, a, this has the effect of taking a photon out of the field at a given time step. There will be some cases where we have 'collapses' where say the photon state is increased! These collapse processes need not correspond solely with energy leaving the system.

When do these collapses occur? The probability of a photodetection in a time step dt is the rate of photons leaving times the time step,

$$P_{\text{coll}} = 2\kappa\langle\psi_c(t)|a^{\dagger}a|\psi_c(t)\rangle dt. \tag{9.15}$$

In general, we will have

$$P_{\text{coll}} = \Gamma\langle\psi_c(t)|C^{\dagger}C|\psi_c(t)\rangle dt. \tag{9.16}$$

We see that the field evolves continuously according to this Hamiltonian, but punctuated by photon emission events which occur with a probability proportional to the photon number.

As an example, for the single cavity field mode the master equation is

$$\dot{\rho} = -i\omega_0[a^\dagger a, \rho] + \kappa(2a\rho a^\dagger - a^\dagger a\rho - \rho a^\dagger a). \tag{9.17}$$

We identify

$$(\mathcal{L} - S)\rho = -i\omega_0[a^\dagger a, \rho] - \kappa[a^\dagger a, \rho]_+ \tag{9.18}$$

$$\mathcal{S}\rho = 2\kappa a\rho a^\dagger. \tag{9.19}$$

with the non-Hermitian Hamiltonian

$$H_{\text{eff}} = \hbar\omega a^\dagger a - i\hbar\kappa a^\dagger a \tag{9.20}$$

At this point, everything we have done is ad hoc; we will formalize it more later. These results can also be obtained by starting from a measurement theory point of view (a là Howard Wiseman); or from the formal results of photon counting developed by Mandel, and Kelly and Kleiner; extended using a Dyson expansion (a là Howard Carmichael). A schematic of the implementation is shown in figure 9.1.

The quantum trajectories are used in simulations as in figure 9.1.

Starting from some initial state, the wavefunction evolves according to the Schrödinger equation

$$\frac{d}{dt}|\psi_{\text{REC}}(t)\rangle = -\frac{i}{\hbar}H_{\text{eff}}|\psi_{\text{REC}}(t)\rangle. \tag{9.21}$$

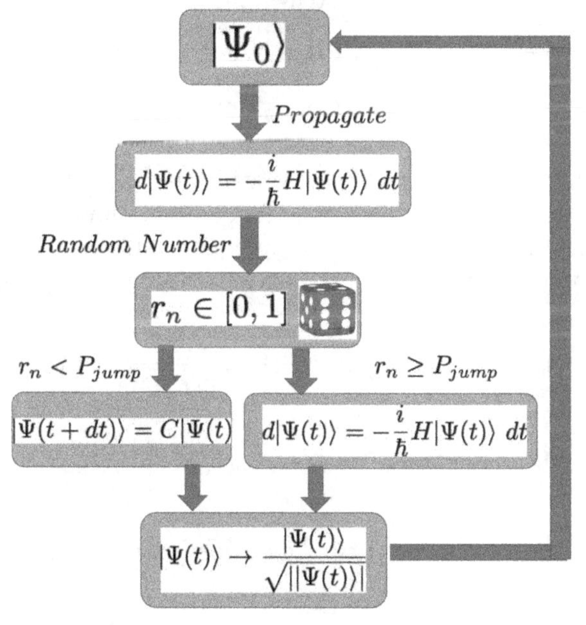

Figure 9.1. Flow chart for quantum trajectory simulations.

This evolution is punctuated by collapses occurring at random times with a probability determined by

$$P_{\text{coll}} = \frac{\langle\psi_{\text{REC}}(t)|\hat{C}^{\dagger}\hat{C}|\psi_{\text{REC}}(t)\rangle}{\langle\psi_{\text{REC}}(t)|\psi_{\text{REC}}(t)\rangle} \tag{9.22}$$

at any given time. The denominator is necessary because the non-Hermitian Hamiltonian does not preserve the norm of the wavefunction. In practice, the wavefunction is numerically integrated and normalized at each time step. The wavefunction $|\psi_c(t)\rangle$ is conditioned on a particular series of collapses having occurred.

Obviously, the state vector changes dramatically upon a detection event, which corresponds to a change in our state of knowledge of the state. The length of the state vector can be drastically changed, and we must normalize it. If there is *not* a click, we also obtain information, but not as much. Here, too we must then normalize the wave function every time step. Mathematically, this results from the fact that a non-Hermitian Hamiltonian leads to a non-unitary propagation, and the norm of the state vector is not preserved.

An ensemble average of conditioned wavefunctions is formally equivalent to the density matrix because

$$\rho = \overline{|\psi_{\text{REC}}(t)\rangle\langle\psi_{\text{REC}}(t)|}. \tag{9.23}$$

Expectation values are calculated from the wavefunction in the usual way and then averaged

$$\langle\hat{O}\rangle = \overline{\langle\psi_{\text{REC}}(t)|\hat{O}|\psi_{\text{REC}}(t)\rangle} \tag{9.24}$$

One can also write the whole process as one differential equation

$$d|\psi_{\text{REC}}\rangle = \frac{-\imath}{\hbar}H_{\text{eff}}|\psi\rangle dt + \left(\sqrt{\Gamma}C - 1\right)|\psi_{\text{REC}}\rangle dN \tag{9.25}$$

where the increment dN satisfies $dN^2 = dN$, which is satisfied by $dN = 0, 1$; that is, either the jump happens or it does not. After every time step, one still has to normalize the wave function. We can interpret the normalization process into our equations directly, which results in a nonlinear Schrödinger equation

$$\begin{aligned} d|\psi_{\text{REC}}\rangle &= \frac{-i}{\hbar}(H_{\text{eff}} + \imath\hbar\Gamma\langle C^{\dagger}C\rangle)|\psi\rangle dt + \left(\frac{2\sqrt{\Gamma}C}{\sqrt{2\Gamma\langle C^{\dagger}C\rangle}} - 1\right)|\psi\rangle dN \\ &= \frac{-\imath}{\hbar}(H_S - i\hbar\Gamma C^{\dagger}C + \imath\hbar\Gamma\langle C^{\dagger}C\rangle)|\psi\rangle dt + \left(\frac{\sqrt{\Gamma}C}{\sqrt{\langle C^{\dagger}C\rangle}} - 1\right)|\psi\rangle dN \end{aligned} \tag{9.26}$$

The nonlinearity stems from the part of the equation involving $\langle C^{\dagger}C\rangle$ multiplying $d|\psi\rangle$.

It can be shown that the ensemble average of conditional density matrices is equivalent to the system's real density matrix:

$$\rho(t) = \overline{|\psi_c(t)\rangle\langle\psi_c(t)|}. \tag{9.27}$$

If we unravel this equation as

$$S\rho = 2\kappa a\rho a^\dagger, \tag{9.28}$$

$$(L - S)\rho = -i\omega_0[a^\dagger a, \rho] - \kappa[a^\dagger a, \rho]_+, \tag{9.29}$$

we will have adopted a formalism based upon a simulated photon counting experiment, where the detector is modeled as an absorptive device. Thus, to calculate the system's photon statistics, we simple employ the same QTF simulation presented earlier and keep a record of collapse event times. By making a histogram of photon absorption delay times, we can characterize the transmitted field's dynamics in precisely the same manner as is done experimentally.

What if there is more than one collapse operator? Let us say there are two, with probabilities P_1 and P_2, both small numbers due mainly to dt. Then we perform collapse 1 if the random number falls between 0 and P_1, collapse 2 if the random number falls between P_1 and P_2, and no collapse if the random number is larger than $P_1 + P_2$. There are other protocols that can be used.

To mathematically formalize this approach, we first *unravel* the master equation by writing equation (9.1) as

$$\dot{\rho} = (\mathcal{L} - S)\rho + S\rho, \tag{9.30}$$

where $S\rho$ contains all terms of form $\hat{O}^\dagger\rho\hat{O}$. These $\hat{O}$ operators are those that model collapse events for the system, although in actuality we can choose *any* operator for $\mathcal{S}$. We express the remaining terms as

$$(L - S)\rho = -\frac{i}{\hbar}[H_S, \rho] + [H_D, \rho]_+, \tag{9.31}$$

where $[\hat{O}, \rho]_+ = \hat{O}\rho + \rho\hat{O}$, and H_D contains the dissipative terms not included in $S\rho$. Equation (9.31) can only be utilized if the master equation is unravelling in modeling a direct detection process. The Hamiltonian that determines the continuous evolution of the pure states is not Hermitian, and is given by

$$\hat{H} = H_s + i\hbar H_D, \tag{9.32}$$

where H_S is the Hermitian Hamiltonian for the system, not accounting for dissipation processes.

We shall see in the next section how this approach allows us to gain more intuition about open systems; going beyond noticing that off-diagonal density matrix elements are nonzero and that there is some type of quantum coherence. Beyond that, what are the advantages of this approach? Well the density matrix involves $N(N-1)/2 \sim N^2$ elements, after taking into account the hermiticity of the density matrix, and the condition that $\mathrm{Tr}(\rho) = 1$. So as our system size increases, the number of equations that must be solved increases quadratically. In the quantum trajectory approach, one has to evolve N equations, a significant savings in space requirements. Of course, the downside is that we must solve these N equations many times in order to perform an ensemble average; the density matrix/quantum regression approach is generally more efficient for crunching numbers.

The unraveling we have examined here involves the choice $\mathcal{S} = \Gamma C$ where C is an annihilation operator of the system, and the unraveling is usually referred to as the direct detection version of quantum trajectories.

The quantum trajectories are used in simulations as follows as in figure 9.1.

9.1 Some examples

Let us consider the example of the damped two-level atom, for which the master equation is

$$\begin{aligned}\dot{\rho} = &-\frac{i}{2}\omega_A[\sigma_z, \rho] + \frac{\gamma}{2}(\bar{n} + 1)(2\sigma_-\rho\sigma_+ - \sigma_+\sigma_-\rho - \rho\sigma_+\sigma_-) \\ &+ \frac{\gamma}{2}\bar{n}(2\sigma_+\rho\sigma_- - \sigma_-\sigma_+\rho - \rho\sigma_-\sigma_+),\end{aligned} \tag{9.33}$$

where γ is the rate of spontaneous emission. To simplify our example, let us assume that the thermal photon number $\bar{n}$ is equal to zero; this is a valid approximation in many experimental regimes, particularly for optical transitions, and will not qualitatively change the form of our example. So, the master equation for our damped two-level atom is now written as

$$\dot{\rho} = -\frac{i}{2}\omega_A[\sigma_z, \rho] + \frac{\gamma}{2}(2\sigma_-\rho\sigma_+ - \sigma_+\sigma_-\rho - \rho\sigma_+\sigma_-). \tag{9.34}$$

We can immediately recognize $\frac{\gamma}{2}(2\sigma_-\rho\sigma_+)$ as a collapse term of the form $\hat{O}^{\dagger}\rho\hat{O}$, and thus we define $S\rho$ to be

$$\begin{aligned}S\rho &= \gamma\sigma_-\rho\sigma_+ \\ &= \sqrt{\gamma}\sigma_-|\psi_{\text{REC}}\rangle\langle\psi_{\text{REC}}|\sqrt{\gamma}\sigma_+.\end{aligned} \tag{9.35}$$

Thus,

$$(L - S)\rho = -\frac{i}{2}\omega_A[\sigma_z, \rho] - \frac{\gamma}{2}[\sigma_+\sigma_-, \rho]_+. \tag{9.36}$$

with effective non-Hermitian Hamiltonian

$$H_{\text{eff}} = \frac{\hbar\omega}{2}\sigma_z - \frac{i\hbar\gamma}{2}\sigma_+\sigma_- \tag{9.37}$$

Equation (9.35) identifies $\sqrt{\gamma}\sigma_-$ as the collapse operator for the system; this follows from our model of radiative damping for the two-level atom, as a transition from the excited state to the ground state will radiate a photon into the environment. With the identification of the appropriate collapse operator, we can now evaluate the probability of a collapse event occurring $P_{\text{coll}}(t)$ during any random time interval $t + dt$.

$$P_{\text{coll}}(t) = \text{tr}[S\rho_{\text{REC}}(t)]dt = \gamma\frac{\langle\psi_{\text{REC}}(t)|\sigma_+\sigma_-|\psi_{\text{REC}}(t)\rangle}{\langle\psi_{\text{REC}}(t)|\psi_{\text{REC}}(t)\rangle}dt, \tag{9.38}$$

where the subscript c denotes the use of the conditional wavefunction and density matrix for the system. The denominator of equation (9.38) is needed because the non-Hermitian Hamiltonian does not generally preserve the normalization of the conditional wavefunction. With these fundamental procedures and details explained, we can now work through the typical operation of a QTF simulation carried out on the computer. First, we consider an ensemble of systems each in some common initial state; for the purpose of this explanation let us follow one such system, and its conditional wavefunction $|\psi_c(t)\rangle$, through the simulation. The systems undergo evolution as governed by the Schrödinger equation:

$$i\hbar|\dot{\psi}_c(t)\rangle = H|\psi_c(t)\rangle, \tag{9.39}$$

where H is the Hamiltonian of equation (9.32). At each time-step (t_n) of the program, a random number (r_n) in the interval [0, 1] is generated and compared to the collapse probability as evaluated in equation (9.38). One of two conditions can be satisfied at this point, determining the evolution of the system for the time-step. If $r_n \geqslant P_{\text{coll}}(t_n)$, then the system will experience a collapse and the evolved conditional wavefunction at time t_{n+1} will be given by

$$|\psi_{rec}(t_{n+1})\rangle = \frac{\sigma_-|\psi_{\text{REC}}(t_n)\rangle}{\sqrt{\langle\psi_{rec}(t_n)|\sigma_+\sigma_-||\psi_{rec}(t_n)\rangle}}. \tag{9.40}$$

If $r_n < p_c(t_n)$, then the system will undergo continuous Hamiltonian evolution for the duration of the time-step, giving the conditional wavefunction to be

$$|\psi_{\text{REC}}(t_{n+1})\rangle = \frac{e^{-(i/\hbar)H\Delta t}|\psi_{\text{REC}}(t_n)\rangle}{\sqrt{\langle\psi_{\text{REC}}(t_n)|e^{(i/\hbar)(H^\dagger - H)\Delta t}|\psi_{\text{REC}}(t_n)\rangle}}. \tag{9.41}$$

Notice that the evolved conditional wavefunctions are to be normalized at each time-step.

For our simulation, we have

$$H_{\text{eff}} = \hbar\omega_0\sigma_z - \imath\hbar\sigma_+\sigma_- \tag{9.42}$$

$$|\psi\rangle = C_g|g\rangle + C_e|e\rangle \tag{9.43}$$

with evolution equations

$$\dot{C}_g = i\omega_0 C_g \tag{9.44}$$

$$\dot{C}_e = -i\omega_0 C_e - \frac{\gamma}{2} C_e \tag{9.45}$$

A simulation for an atom initially in the excited state is shown below in figure 9.2. The excited state probability stays at 1 until there is a jump, and then the excited state is then 0. There is no evolution between jumps here. The excited state probability amplitude decays via the non-Hermitian $-i\hbar\gamma$ term, but then when we normalize the wave function, that cancels that out.

As one increases the number of trajectories averaged over, we see convergence to the full master equation result. We see N jumps in general for N trajectories, although they may occur simultaneously (figure 9.3). The master equation result obtains in the $N \rightarrow \infty$ limit, but sometimes 10–100 is quite enough for simple systems.

Now consider the single mode cavity, described by

$$|\psi\rangle = \sum_n C_n |n\rangle \tag{9.46}$$

$$H_{\text{eff}} = \hbar\omega_0 a^\dagger a - \imath\hbar\kappa a^\dagger a \tag{9.47}$$

The probability for a jump at any time step is $P_c = 2\kappa\langle\psi|a^\dagger a|\psi\rangle dt$, and the probability amplitudes have a nonjump evolution described by

$$\dot{C}_n = -(\imath\omega_0 + 2\kappa)nC_n \tag{9.48}$$

with solution

$$C_n(t) = \exp(-(\imath\omega_0 + 2\kappa)n)C_n \tag{9.49}$$

There is exponential decay of these amplitudes, reflecting the fact that the *lack* of a detection event tells us something about the relative probability of the various states. But as in the spontaneous emission example, that is countered by the decrease in the

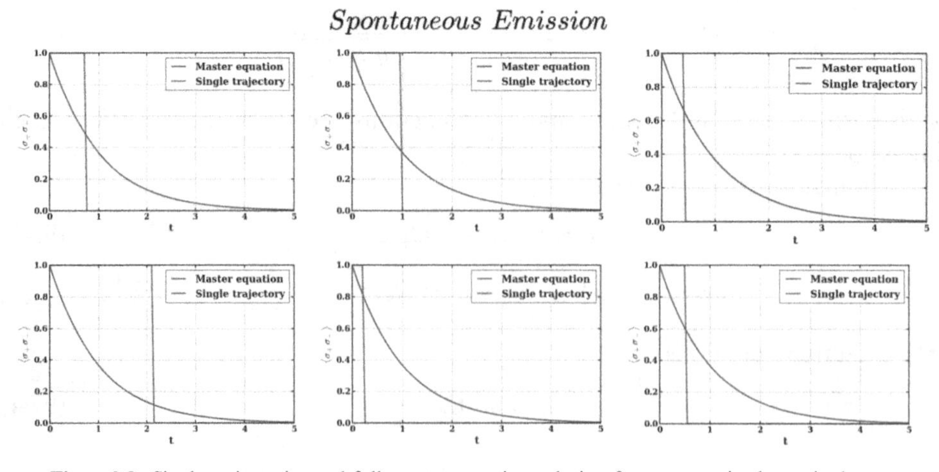

Figure 9.2. Single trajectories and full master equation solution for an atom in the excited state.

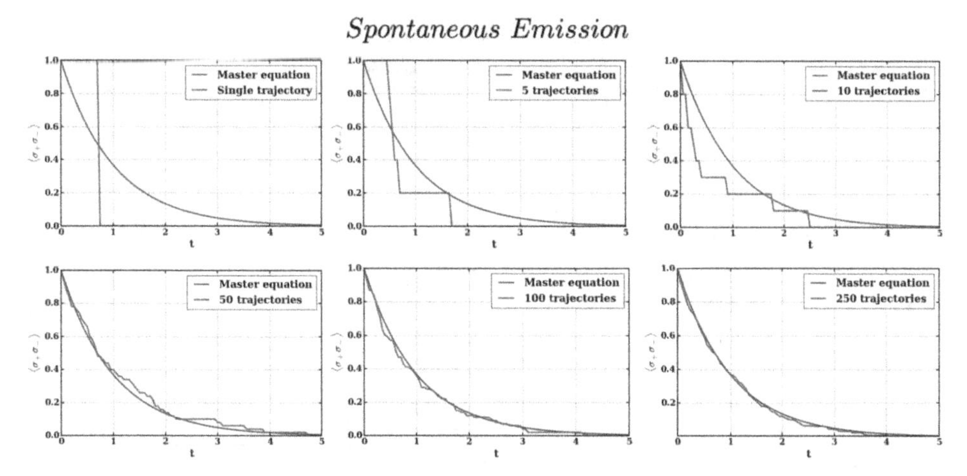

Figure 9.3. Spontaneous emission simulations for increasing number of trajectories averaged over.

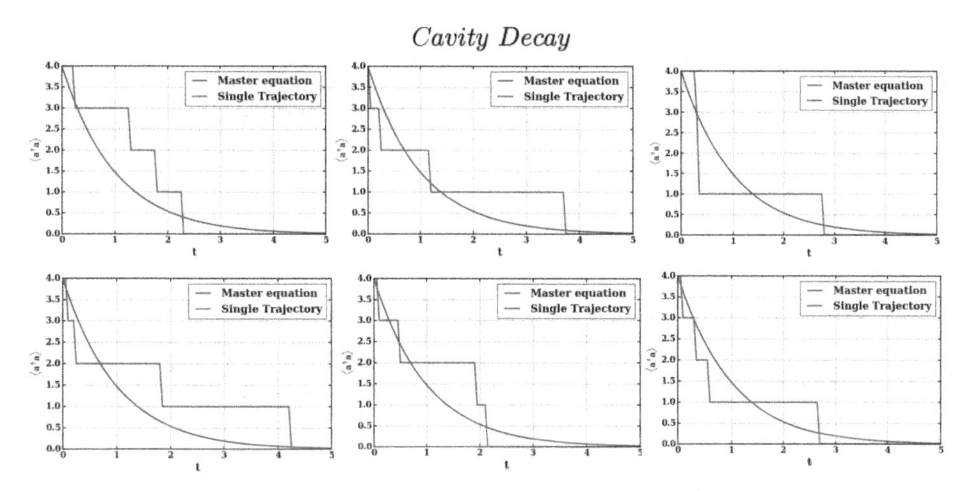

Figure 9.4. Single trajectories and master equation solution for a cavity initially in a Fock state with four photons.

norm of the state. A sample trajectory is shown in figure 9.4, where we start with four photons in the cavity and plot $\langle a^{\dagger}(t)a(t)\rangle$. We see four jumps at random times, eventually emptying the cavity.

Next, we examine the effect of increasing the number of trajectories, and see a fairly rapid convergence to the full master equation result.

In the case of a highly excited cavity (figure 9.5), the effect of the jumps becomes less noticeable, and a single trajectory can well approximate the master equation solution as we show in figure 9.6.

A particularly interesting case is that of a cavity prepared in a coherent state. There is *no* collapse in the sense that the collapse operator does *not* change the state

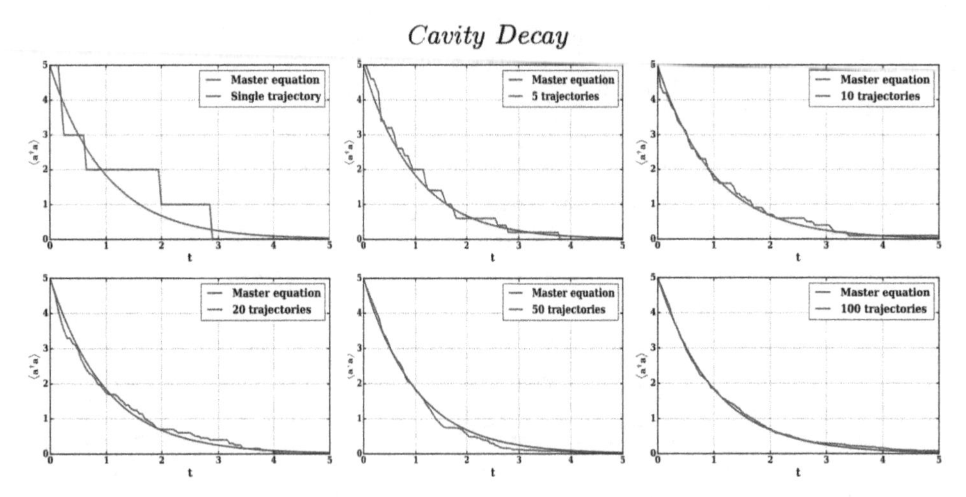

Figure 9.5. Cavity decay simulations for increasing number of trajectories averaged over.

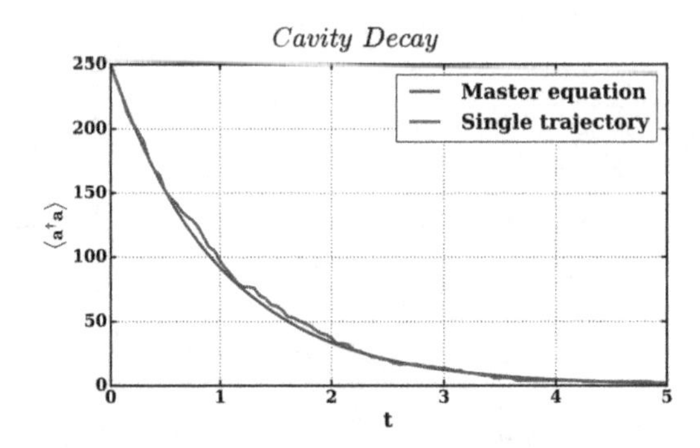

Figure 9.6. Single trajectory for decay of a cavity initially with 250 photons in it.

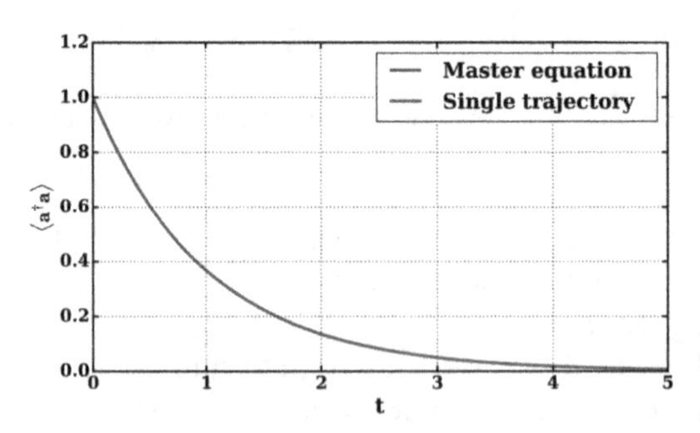

Figure 9.7. Single trajectory for a cavity in a coherent state with $\alpha = 1$.

of the system, $a|\alpha\rangle = \alpha|\alpha\rangle \rightarrow |\alpha\rangle$, where the latter relation holds after normalization. The non-jump evolution gives us

$$\begin{aligned} |\psi(t)\rangle &= e^{-2\kappa\alpha t}|\psi(0)\rangle \\ &= |\alpha_0 e^{-\kappa t}\rangle \end{aligned} \tag{9.50}$$

This is another indication of the classical nature of the coherent state! See figure 9.7 for the case of a coherent state with one photon on average.

How about if we start with a superposition state $(1/\sqrt{2})(|2\rangle + |0\rangle)$? An example trajectory and ensemble average is shown in figure 9.8.

Here, we see exponential decay and upward collapses. If a photon is detected, it must have come from the $n = 2$ part of the state, leaving the cavity in $n = 1$, which is higher than the master equation solution; the collapse operator annihilates the vacuum. Occasionally, there are no collapses as the trajectory decays faster than the master equation solution. Of course when averaged over many trajectories, this balances out to give the master equation solution as it must, as shown in figure 9.9.

Another case of interest is a cavity with no photons coupled to a thermal field with $\langle n\rangle = 5$. We see a mixture of upward jumps, with some downward, with lots of noise for single trajectories as in figure 9.10.

And as we expect, as we average over a set of trajectories the master equation result eventually obtains as the number of trajectories n_{traj} is increased, as in figure 9.11.

Let us now consider an optical parametric oscillator. This is a device with a $\chi^{(2)}$ crystal that is pumped externally and results in pairs of photons being generated in the cavity. The Hamiltonian is

$$H = \Omega(a^2 + a^{\dagger 2}) \tag{9.51}$$

We assume no depletion of the pump, and assume we are pumping at a small enough level to avoid gain saturation. For $\Omega = 1$, we see in figure 9.12 that the average photon number rises to a level ϵ below the master equation solution, and then there is an upward jump of $1 + \epsilon$. Then there is decay back to ϵ. For weak pumps, the wave function is dominated by the vacuum, with single and dual photon state amplitudes on the order of ϵ^2 and then higher order terms Upon collapse, the vacuum and single photon states are annihilated, and after normalization the

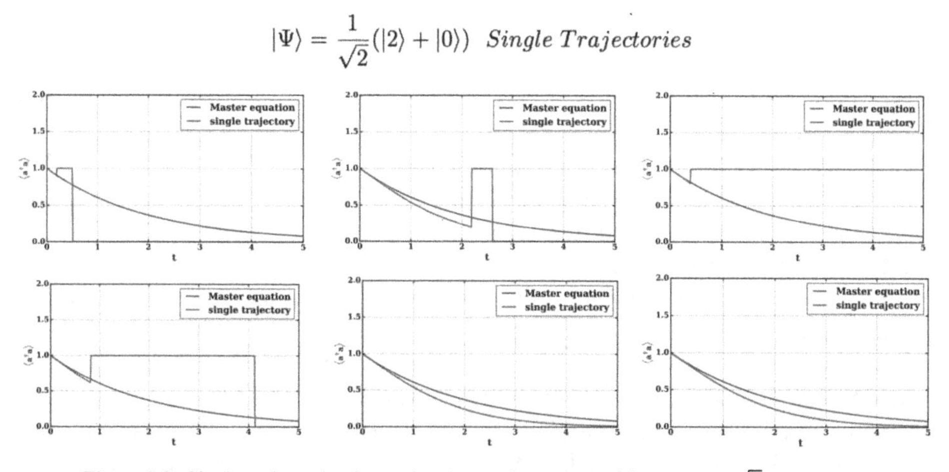

Figure 9.8. Single trajectories for cavity decay of a superposition state $(1/\sqrt{2})(|2\rangle + |0\rangle)$.

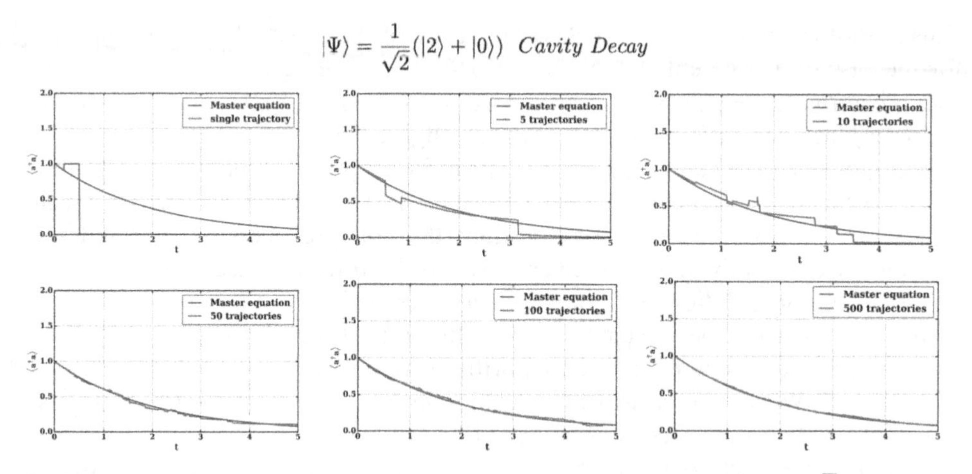

Figure 9.9. Increasing number of trajectories for cavity decay of a superposition state $(1/\sqrt{2})(|2\rangle + |0\rangle)$.

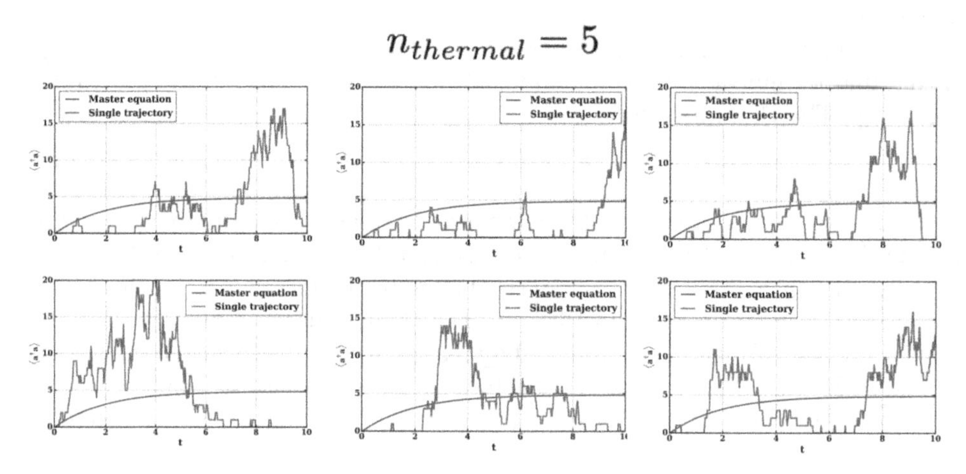

Figure 9.10. Single trajectories for cavity decay of a thermal state.

average photon number is greater than 1, reflecting the now dominant contributions from $a|2\rangle$, and less from $a|3\rangle$, etc.

When one turns up the pump some, we notice a very noisy trace for single trajectories as in figures 9.13 and 9.14. The OPO has a threshold, and we see hints of the large noise near that phase transition.

Returning to the weak pumping case, as we increase the number of trajectories we see many upward jumps initially, but they decrease in magnitude and the master equation result of course obtains.

As our last example, we will consider a two-level atom driven by a classical field, resonance fluorescence. For Rabi frequencies well below the spontaneous emission rate, we see in figure 9.15 for a single trajectory an initial increase in the excited state population, and then a jump back to the ground state, and the process starts anew.

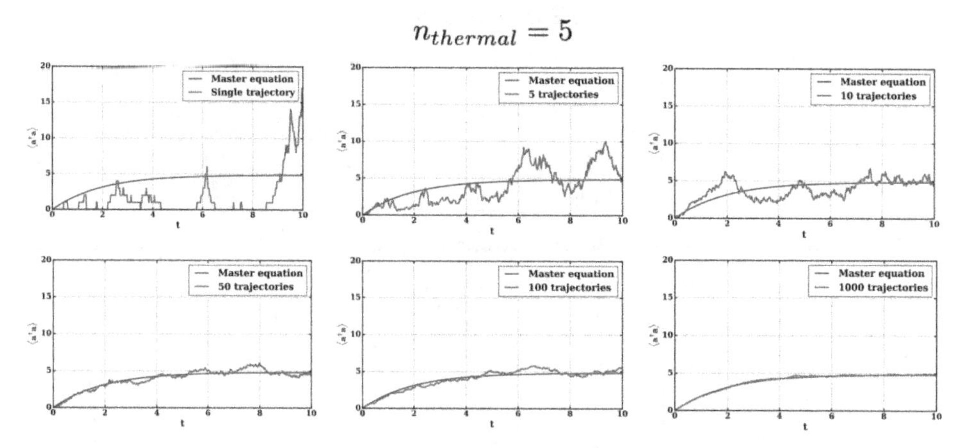

Figure 9.11. Increasing number of trajectories for cavity decay of a thermal state.

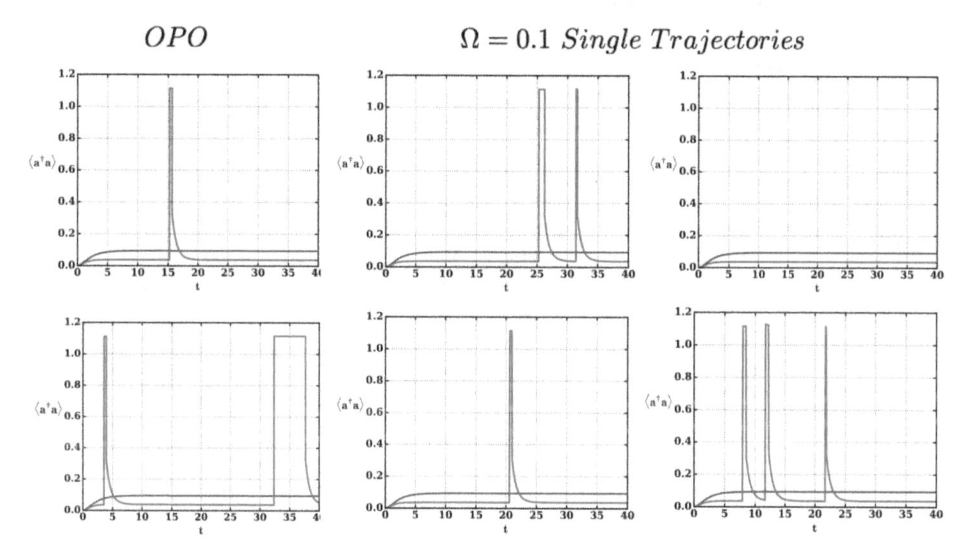

Figure 9.12. Single trajectories for a weakly pumped OPO.

In figure 9.16, we see that increasing the Rabi frequency to twice the spontaneous emission rate results in visible oscillations, followed by jumps.

9.2 And now for a little formality

So, how does this dual-evolution stem from the master equation in a more formal way (figures 9.17 and 9.18).

The formal solution to the master equation is

$$\rho(t) = e^{\mathcal{L}t}\rho(0) \tag{9.52}$$

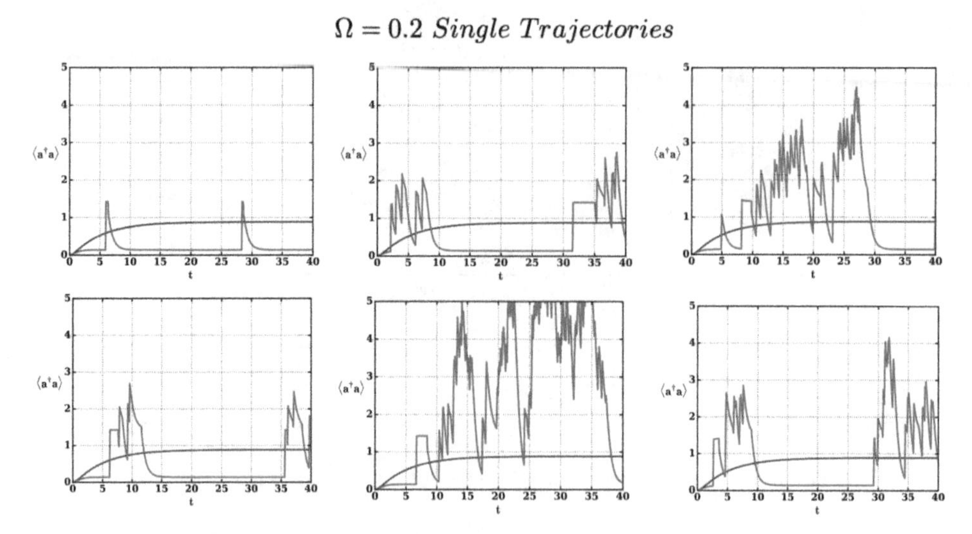

Figure 9.13. Single trajectories for an OPO slightly below threshold.

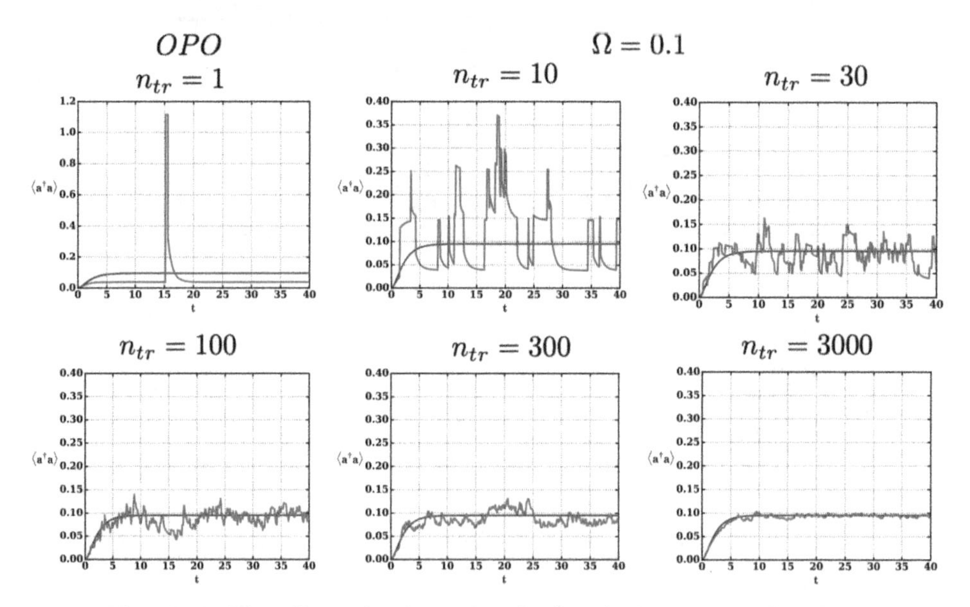

Figure 9.14. Effect of increasing the number of trajectories for a weakly driven OPO.

We use the following identity, known as the Dyson expansion,

$$
\begin{aligned}
e^{\mathcal{L}t} &= e^{[(\mathcal{L}-\mathcal{S})+\mathcal{S}]t} \\
&= \sum_{m=0}^{\infty} \int_0^t dt_m \int_0^{t_m} dt_{m-1} \cdots \int_0^{t_2} dt_1 e^{(\mathcal{L}-\mathcal{S})(t-t_m)}\mathcal{S} \\
&\quad \times e^{(\mathcal{L}-\mathcal{S})(t_m-t_{m-1})}\mathcal{S} \cdots \mathcal{S}e^{(\mathcal{L}-\mathcal{S})t_1}\rho(0)
\end{aligned}
\tag{9.53}
$$

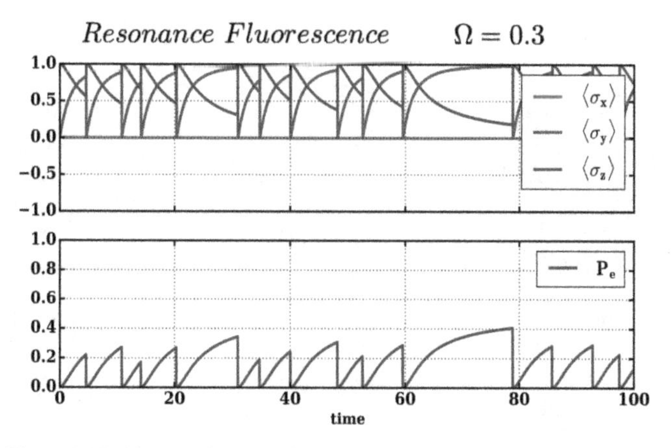

Figure 9.15. Single trajectories for weakly driven resonance fluorescence.

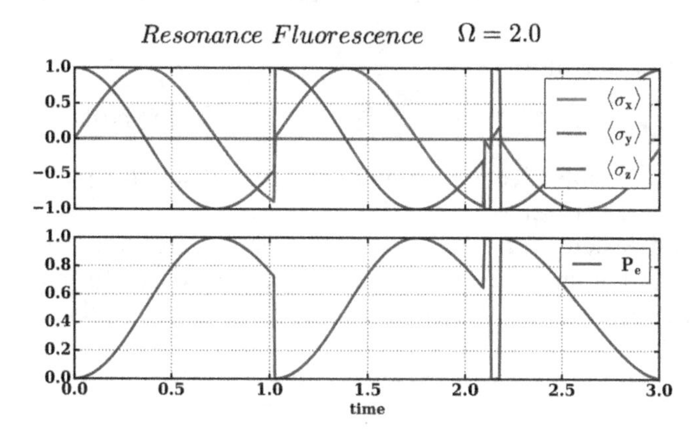

Figure 9.16. Single trajectories for strongly driven resonance fluorescence.

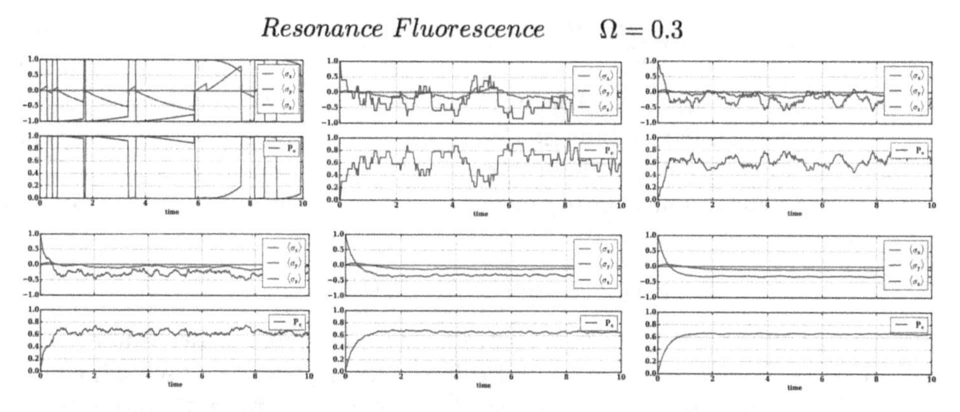

Figure 9.17. Effect of increasing number of trajectories for weakly driven resonance fluorescence.

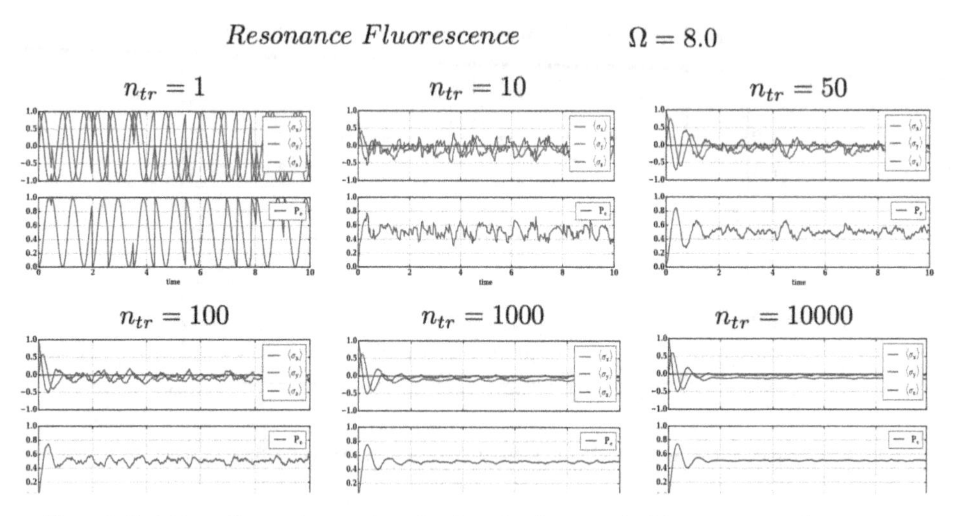

Figure 9.18. Effect of increasing number of trajectories for strongly driven resonance fluorescence.

We now define an operator

$$\bar{\rho}_{\mathrm{REC}} = e^{(\mathcal{L}-\mathcal{S})(t-t_m)}\mathcal{S} \times e^{(\mathcal{L}-\mathcal{S})(t_m-t_{m-1})}\mathcal{S} \cdots \mathcal{S}e^{(\mathcal{L}-\mathcal{S})t_1}\rho(0) \tag{9.54}$$

The subscript REC stems from the notion that this is a density matrix-like object *conditioned* on a given set of *detection* events. If we use $\mathcal{S} = \sqrt{2\kappa}a$, then these detections would be actual photodetector clicks. We are free to use any choice for $\mathcal{S}$; one can optimize the choice depending on what one wishes to measure/calculate. This can be used to define a density operator

$$\rho_{\mathrm{REC}}(t) = \frac{\bar{\rho}_{\mathrm{REC}}}{\mathrm{Tr}[\bar{\rho}_{\mathrm{REC}}]} \tag{9.55}$$

which is properly normalized. We can then write

$$\rho(t) = \sum_{m=0}^{\infty} \int_0^t dt_m \int_0^{t_m} dt_{m-1} \cdots \int_0^{t_2} dt_1 \bar{\rho}_{\mathrm{REC}} \tag{9.56}$$

So, we can see that the 'real' density matrix is a sum over conditioned density matrices, that correspond to different detection records. This is sort of a Feynman diagrammatic version of density matrix theory. It is most useful to make a choice for $\mathcal{S}$ for which the conditioned density operator can be factorized, as in the examples we have seen already,

$$\rho_{\mathrm{REC}}(t) = |\psi_{\mathrm{REC}}(t)\rangle\langle\psi_{\mathrm{REC}}(t)| \tag{9.57}$$

where we can use this decomposition for either the normalized or un-normalized version (the overbarred version or not). Each of these trajectories, or wave functions $|\psi_{\mathrm{REC}}\rangle$ evolve in time via an evolution via $e^{(\mathcal{L}-\mathcal{S})t}$ punctuated by applications of $\mathcal{S}$ at the times $t_1, t_2, \ldots, t_m$, where m is the total number of detection events for a particular trajectory. We use the phrase detection event, but some would use collapse. The collapse term can cause long arguments; it is a collapse in the Bayesian sense of a dramatic change in our knowledge of the system, and not necessarily in the Bohr sense of an actual collapse of a physical system. How do we calculate expectation values?

$$\langle O \rangle = \mathrm{Tr}(\rho O) \tag{9.58}$$

In terms of the trajectory decomposition, we have

$$\begin{aligned} \langle O \rangle &= \mathrm{Tr}(\Sigma |\psi_{\mathrm{REC}}(t)\rangle\langle\psi_{\mathrm{REC}}(t)| O) \\ &= \Sigma\, \mathrm{Tr}(\langle\psi_{\mathrm{REC}}(t)|O|\psi_{\mathrm{REC}}(t)\rangle) \\ &= \Sigma \langle\psi_{\mathrm{REC}}(t)|O|\psi_{\mathrm{REC}}(t)\rangle \end{aligned} \tag{9.59}$$

where we have used the shorthand $\Sigma \equiv \sum_{m=0}^{\infty} \int_0^t dt_m \int_0^{t_m} dt_{m-1} \cdots \int_0^{t_2} dt_1$ which is essentially an ensemble average over all realizations of the trajectories. This is a generalized sum over the time and total number of detection events. The method usually used to calculate these averages is to use an ensemble of trajectories, normalizing the trajectories at each time step, and calculating

$$\frac{1}{N} \sum_{\mathrm{REC}} \langle\psi_{\mathrm{REC}}(t)|O|\psi_{\mathrm{REC}}(t)\rangle \tag{9.60}$$

where N is the number of trajectories simulated in the manner discussed earlier in this chapter. To calculate steady state values of expectation values, one can usually replace the ensemble average with a time average, while mumbling something about ergodicity.

9.3 Homodyne detection and quantum state diffusion

Consider the homodyne detection scheme in figure 7.3. A photodetector would generate a current proportional to the intensity, as

$$i(t) \sim (rE_{LO} + tE_{\mathrm{signal}})^2 \tag{9.61}$$

$$= X^2 + 2XtE_{\mathrm{signal}} + t^2 E_{\mathrm{signal}}^2 \tag{9.62}$$

where we have defined $rE_{LO} \equiv X$. In such a scheme, one does not want to lose much signal, so $t \to 1$, and one would like to have a strong local oscillator such that $X = rE_{LO} \gg tE_{\mathrm{signal}}$. In such a case, the photodetector will fire almost continuously. Each detection tells us something about the combined local oscillator/signal field, but very little about the signal itself. In fact, in a time increment dt that is small compared to $1/f_{\max}$, where $f_{\max}$ is the largest frequency in the signal field that is of interest, there will be a large number of detection events. To utilize the quantum trajectory approach as we have developed it, we would have to go to a time step small compared to the inverse count rate due to the local oscillator, $1/X^2$, which decreases with the intensity of the local oscillator. We do not wish to keep track of every detection anymore, just examine the dynamics of the system on a time scale relevant to it. Also, the collapse operator is no longer one that takes a photon away from the source; that is not the action of the detector in the homodyne detection scheme. We can address these problems by a new choice for the unraveling of our master equation, with

$$\mathcal{S} = x + \sqrt{2\kappa}\, a_{\text{signal}} \tag{9.63}$$

where x is the field of local oscillator multiplied by r, in photon flux units,and we have set $t = 1$ and used the signal field in photon flux units $E_{\text{signal}} \rightarrow \sqrt{2\kappa}\, a$. This choice takes into account that collapses do *not* reflect a photon coming from the source. The other problem, that we would have to use a very small time step, can be avoided by a type of coarse graining. We look at time increments over which the system does not change very much, but during which the number of photocounts is large enough to consider the number of counts to be Poisson-distributed.

This results in

$$\begin{aligned}(\mathcal{L} - S)\rho &= (-i/\hbar)[H_S, \rho] - \kappa a^\dagger a\rho - \kappa\rho a^\dagger a - x^2\rho - x\sqrt{2\kappa}(\rho a^\dagger + a\rho) \\ &= (-i/\hbar)[H_{\text{eff}}, \rho]\end{aligned} \tag{9.64}$$

where we have an effective Hamiltonian

$$H_{\text{eff}} = H_S - i(\sqrt{\kappa}x(a - a^\dagger) + (x + \sqrt{2\kappa}a)(x + \sqrt{2\kappa}a^\dagger)) \tag{9.65}$$

This is then the Hamiltonian used in the no-jump evolution. What about the jumps? Well the probability of a collapse in time dt is given by

$$P_{\text{coll}} = 2\kappa\langle(x^2 + \sqrt{2\kappa}x(a + a^\dagger) + 2\kappa a^\dagger a)\rangle dt \tag{9.66}$$

As we can see, to resolve the jumps one must have $dt \ll 1/x^2$, but we do not wish to resolve all of these clicks! To see how to proceed we return to the nonlinear Schrödinger equation that contains both jumps and jump-free evolution

$$d|\psi\rangle = \frac{-i}{\hbar}(H_S - i\hbar\Gamma C^\dagger C - i\hbar\Gamma\langle C^\dagger C\rangle)|\psi\rangle dt + \left(\frac{C}{\langle C^\dagger C\rangle - 1}\right)|\psi\rangle dN \tag{9.67}$$

where we must use $C = x + \sqrt{2\kappa}a$. We now wish to do the coarse graining; we begin by considering the number of jumps in a time step dt; we no longer wish to look at a case where the jump happens or not, with small probability but consider a situation where there are many detection events in time interval dt. We start with

$$\bar{dN} = (x^2 + x\sqrt{2\kappa}\langle a - a^\dagger\rangle + 2\kappa\langle a^\dagger a\rangle)dt \tag{9.68}$$

where we will drop the last term, as we are only interested in keeping the leading term in the expression that involves the source, and take the limit $x \gg \langle a^\dagger a\rangle$. Now, in any time interval dt, we will not always obtain the same number of counts; there will be fluctuations. If the number of counts is large enough, they will be Poisson distributed about the mean. This is not just an appeal to the mean value theorem, it is the strict result of applying the quantum theory of detection, according to the work of Mandel, and Kelly and Kleiner. In this case, we may write down a 'noisy' count operator

$$dN = (x^2 + \sqrt{2\kappa}x(a + a^\dagger))dt + xdW \tag{9.69}$$

where we have kept the leading noise term, and dW is a Weiner process defined by

$$\langle dW \rangle = 0 \tag{9.70}$$

$$\langle dW^2 \rangle = dt \tag{9.71}$$

This is the type of noise one finds in treatments of Brownian motion, where the mean displacement of an object under a random force is zero, but the root-mean-square displacement is proportional to $\sqrt{dt}$. Operationally, to implement this noise process, one draws a random number from a Gaussian distribution with a unit standard deviation, and multiplies it by $\sqrt{dt}$. Now, we have neglected terms of lower order in x so far, for the sake of consistency, we must expand equation (9.67) in powers of x. From the jump part of equation (9.67), we obtain terms on the order of dt and $dW \sim \sqrt{dt}$; the non-jump evolution gives us terms proportional to dt. After quite a bit of algebra we find

$$\begin{aligned} d|\psi\rangle = & \left[-\frac{i}{\hbar}H_S - \kappa(a^\dagger a - 2\langle a_x\rangle a + \langle a_x\rangle^2)\right]|\psi\rangle dt \\ & + 2\kappa(a - \langle a_x\rangle)|\psi\rangle dW \end{aligned} \tag{9.72}$$

where we have the quadrature operator

$$a_x = \frac{1}{2}(a + a^\dagger) \tag{9.73}$$

This equation was obtained by considering a choice of superoperator $\mathcal{S}$ appropriate for homodyne detection. Carmichael has obtained the same equation by direct calculation, starting with a Hamiltonian for the source and local oscillator; Wiseman has obtained this formula by considering a unitary transformation motivated by considerations of homodyne detection. There are other forms in the literature, particularly that of quantum state diffusion, introduced by Gisin and Percival,

$$\begin{aligned} d|\psi\rangle = & \left[-\frac{i}{\hbar}H_S - \kappa(a^\dagger a - 2\langle a^\dagger\rangle a + \langle a\rangle\langle a^\dagger\rangle)\right]|\psi\rangle dt \\ & + 2\kappa(a - \langle a\rangle)|\psi\rangle dW \end{aligned} \tag{9.74}$$

which can be obtained from our equation by considering a transformation $|\psi(t)\rangle \to exp[i\phi(t)]|\psi(t)\rangle$, with

$$d\phi = 2\kappa\langle a_x\rangle\langle a_y\rangle dt + \sqrt{2\kappa}\langle a_y\rangle dW \tag{9.75}$$

with the other quadrature operator

$$a_y = \frac{1}{2i}(a - a^\dagger) \tag{9.76}$$

This phase is time-varying, but is an overall phase and so can be ignored; either version of the stochastic Schrödinger equation will work. In all of these versions, we have a state vector that evolves in a continuous way, albeit with a stochastic element. If we preserve the normalization in the equation for $d|\psi\rangle$, we also have a nonlinearity as before. Physically, this does correspond to our state of knowledge of the system *if* we monitored it with a homodyne detector; the unravelling can also be used to calculate whatever we need for this system although for photon statistics it is much more difficult than using the direct detection unraveling.

9.4 Two-time averages

How do we use the quantum trajectory approach to calculate two-time averages needed for correlation functions? In the density matrix formalism, we were able to use the quantum regression theorem, where we solved for steady state density matrix elements. These were then used as initial conditions to evolve the two-time averages, using the same Liouvillian. How do we proceed here? Let us start with the example of

$$\eta = \langle a^\dagger(0)a^\dagger(\tau)a(\tau)a(0)\rangle \tag{9.77}$$

This is the numerator of $g^{(2)}(\tau)$, the denominator being $\langle n\rangle^2 = \langle a^\dagger a\rangle_{SS}^2$. In this case we have a symmetric two-time average, in that we have an equal number of creation and annihilation operators acting at similar time. We can define a quantum state $|\psi_C\rangle = a|\psi(t)\rangle$, and we can then write

$$\eta = \langle\psi_C|a^\dagger(\tau)a(\tau)|\psi_C\rangle \tag{9.78}$$

so that our answer is just the inner product of a state, where that state is evolved for a time τ after a collapse. If the system has a steady state, one finds the steady state wave function, applies the collapse operator and evolves. The further evolution can contain further jumps of course. If there is no steady state, but the system is stationary, the choice of initial time will change the trajectories, but *not* the ensemble average.

For

$$\langle a^\dagger(\tau)a(0)\rangle \tag{9.79}$$

we must evolve *two* stochastic wave functions.

$$|A(t)\rangle = U(t)a|\psi(0)\rangle \tag{9.80}$$

$$|B(t)\rangle = aU(t)|\psi(0)\rangle \tag{9.81}$$

The inner product of these two leaves us with the desired result

$$\langle\psi(0)|U^\dagger(\tau)a^\dagger(\tau)a(0)U(\tau)|\psi(0)\rangle \tag{9.82}$$

9.5 Somc final thoughts on trajectories

We mention that there are alternative algorithms available. One common one is to use the waiting time distribution for a jump

$$W(t) = e^{-\gamma t} \tag{9.83}$$

and then select a random number in the interval [0, 1] and evolve until the norm of the state equals that random number, and then apply a jump. In cases where the waiting time distribution is known, analytic answers are sometimes possible.

Finally, what element of reality, if any, can we assign these trajectories? The best we can say is that with a complete measurement record, we can *retrodict* what one member of an ensemble would have been doing. Different unravelling/measurement schemes yield different results for that of course. So perhaps we can say that we play dice with the world, but ones that are loaded based on our measurement record. An individual system is of course not deterministic.

References

[1] Carmichael H J 1993 *An Open Systems Approach to Quantum Optics* (Berlin: Springer)

[2] Carmichael H J 2007 *Statistical Methods in Quantum Optics. 2: Non-Classical FieldsTheoretical and Mathematical Physics* (Berlin: Springer)

[3] Carmichael H J 1999 *Statistical Methods in Quantum Optics. 1: Master equations and Fokker-Planck equationsTheoretical and Mathematical Physics* (Berlin: Springer)

[4] Gisin N and Percival I C 1992 The quantum-state diffusion model applied to open systems *J. Phys. A: Math. Gen.* **25** 5677

[5] Jacobs K and Steck D 2006 A straightforward introduction to continuous quantum measurement *Contemp. Phys.* **47** 279

[6] Mølmer K, Castin Y and Dalibard J 1993 Monte Carlo wave-function method in quantum optics *J. Opt. Soc. Am. B* **10** 524

[7] van Dorsselaer F E and Nienhuis G 2000 Quantum trajectories *J. Opt. B: Quantum Semiclassical Opt* **2** r25

[8] Wiseman H 1996 Quantum trajectories and quantum measurement theory *Quant. Semiclassical Opt.: J. Eur. Opt. Soc.* B **8** 205

IOP Publishing

Chapter 10

Quasiprobability distributions

10.1 Glauber–Sudarshan P representation

As the coherent states form a basis set, albeit overcomplete, we may write the density matrix as [2, 8]

$$\rho = \int P(\alpha)|\alpha\rangle\langle\alpha|d^2\alpha \tag{10.1}$$

where $P(\alpha)$ is a probability distribution. This just says that we can represent the density matrix as a weighted sum over coherent state density matrices. We anticipate where the quantum weirdness is manifest here; the coherent states are essentially classical fields with quantum noise. It would seem that we will not see quantum effects, but we know there are systems that display behavior that can only be described by a quantum field. The rub will turn out to be that $P(\alpha)$ cannot have the properties of a classical distribution function, positive definiteness for one. How can we have squeezed fluctuations in a sum of vacuum fluctuations? Only if there is destructive interference, which would be described by negative values of $P(\alpha)$ for some α. We do require that

$$\int P(\alpha)d^2\alpha = 1 \tag{10.2}$$

How may we use this distribution to calculate expectation values? Consider

$$\langle a^\dagger a\rangle = \mathrm{Tr}(\rho a^\dagger a) \tag{10.3}$$

$$= \mathrm{Tr}\Big(\int P(\alpha)a^\dagger a|\alpha\rangle\langle\alpha|d^2\alpha\Big) \tag{10.4}$$

$$= \int (\mathrm{Tr}(P(\alpha)|\alpha|^2|\alpha\rangle\langle\alpha|)d^2\alpha \tag{10.5}$$

$$= \int (P(\alpha)|\alpha|^2 \operatorname{Tr}(\langle\alpha|\alpha\rangle) d^2\alpha \tag{10.6}$$

$$= \int P(\alpha)|\alpha|^2 d^2\alpha \tag{10.7}$$

where we have used $\langle\alpha|\alpha\rangle = 1$, $a|\alpha\rangle = \alpha|\alpha\rangle$, and the cyclic property of the trace. This can be generalized to

$$\langle a^{\dagger n} a^m \rangle = \int P(\alpha)\alpha^{*n}\alpha^m d^2\alpha \tag{10.8}$$

Averages of this form are referred to as *normal ordering*, where all the annihilation operators are to the left acting on the bra, and the creation operators act to the right on the ket. It is possible to express averages in this form by using the proper commutation relations, for example

$$\langle aa^\dagger \rangle = \langle a^\dagger a \rangle + 1 \tag{10.9}$$

This is only possible because the coherent states are overcomplete. Glauber initially investigated

$$\rho = \frac{1}{\pi^2} \int d^2\alpha \int d^2\beta |\alpha\rangle\langle\alpha|\rho|\beta\rangle\langle\beta| \tag{10.10}$$

$$= \int d^2\alpha \int d^2\beta R(\alpha, \beta)\rho_{\alpha,\beta} \tag{10.11}$$

Notice the P distribution only relies on diagonal matrix elements of ρ in the coherent state basis.

Before turning to examples, let us consider the normally ordered characteristic function

$$\chi_N = \operatorname{tr}(\rho e^{i\beta a^\dagger} e^{i\beta^* a}) \tag{10.12}$$

One can use this to obtain all normally ordered averages via

$$\langle a^{\dagger n} a^m \rangle = \frac{\partial^{m+n}\chi_N}{\partial^m(i\beta)\partial^n(i\beta^*)}\bigg|_{\beta=\beta^*=0} \tag{10.13}$$

Hence, there must be some relationship between this characteristic function and $P(\alpha)$, they are in fact Fourier pairs

$$P(\alpha) = \frac{1}{\pi^2} \int \chi_N(\beta, \beta^*) e^{i\beta a^\dagger} e^{i\beta^* a} \tag{10.14}$$

and

$$\chi_N(\alpha) = \frac{1}{\pi^2} \int P(\beta, \beta^*) e^{-i\beta a^\dagger} e^{-i\beta^* a} \tag{10.15}$$

We can treat α as a complex number with real and imaginary parts, or α and α^* as two independent variables, and we will have use of that again later.

The easiest state to find the P distribution for is a coherent state, it is merely

$$P_{\alpha_0}(\alpha) = \delta(\alpha - \alpha_0) \tag{10.16}$$

Another simple case is that of a thermal state with

$$\bar{n} = \langle n \rangle = \frac{1}{e^{\hbar\omega/kT} - 1} \tag{10.17}$$

There are only diagonal elements of the density matrix for a thermal system, and one finds

$$P_{\text{thermal}}(\alpha) = \frac{1}{\pi\bar{n}} e^{-|\alpha|^2/\bar{n}} \tag{10.18}$$

a simple Gausssian.

Recall that the density operator for a thermal state is

$$\rho_{\text{thermal}} = \frac{1}{Z} e^{-\beta a^\dagger a \hbar\omega} \tag{10.19}$$

with $Z = \text{Tr}(e^{-\beta a^\dagger a \hbar\omega})$. Inserting a coherent state identity, we find m

$$\rho_{\text{thermal}} = \frac{1}{\pi Z} \int d^2\alpha e^{-\beta a^\dagger a \hbar\omega/2} |\alpha\rangle\langle\alpha| e^{-\beta a^\dagger a \hbar\omega/2} \tag{10.20}$$

$$= \frac{1}{\pi Z} \int d^2\alpha e^{-\beta a^\dagger a \hbar\omega/2} |\alpha\rangle\langle\alpha| e^{-|\alpha|^2/\bar{n}} \tag{10.21}$$

The P distribution for a Fock state is a highly singular function. In fact, nonclassical states have this type of behavior. In fact, negative values of $P(\alpha)$ are associated with nonclassical states. Defining $\Delta X = X - \langle X \rangle$ for the X quadrature, we have

$$|\Delta X|^2 = \frac{1}{4} \int P(\alpha) |\Delta\alpha^* + \Delta\alpha|^2 d^2\alpha \tag{10.22}$$

For this to be below 1/4, $P(\alpha)$ must be negative for some values of α.

We will not derive the formula, but we have

$$P_{\text{Fock}}(\alpha) = \frac{e^{|\alpha|^2}}{n!} \frac{\partial^{2n}}{\partial^n\alpha \partial^n\alpha^*} \delta^2(\alpha) \tag{10.23}$$

an nth-order derivative of a two-dimensional δ function. These are called generalized functions. As with a δ function, they are not really functions in the mathematical sense, but make sense under an integral where they play the role of a distribution. A physicist, Dirac, invented them, so we reserve the right to be lazy and refer to them as functions, while keeping their true nature in our minds. It makes conversations easier really, just a convention. It is often more convenient (although still not *convenient*) to use this in a polar form

$$P_{\text{Fock}=l}(\alpha) = \frac{1}{2\pi r}\frac{l!}{(2l)!}e^{r^2}\frac{\partial^{2l}}{\partial^{2l} r} \tag{10.24}$$

We can use these forms for some calculations, for example, the average photon number. Consider the case $|l = 1\rangle$

$$\langle a^\dagger a\rangle = \frac{1}{2}\int r^2 e^{r^2}\frac{\partial^2}{\partial^2 r}\delta(r) \tag{10.25}$$

So, only two derivatives of a δ function.

We use the identity

$$\int f(x)\delta'(x)dx = -f'(0) \tag{10.26}$$

twice, as well as

$$r^2 e^{r^2} = \frac{1}{4}\left(\frac{\partial^2}{\partial^2 r} - 2\right)e^{r^2} \tag{10.27}$$

and find

$$\langle a^\dagger a\rangle_1 = \frac{1}{2}\int r^2 e^{r^2}\frac{\partial^2}{\partial^2 r}\delta(r) \tag{10.28}$$

Using

$$\frac{\partial^2}{\partial^2 r}e^{r^2} = (2 + 4r^2)e^{r^2} \tag{10.29}$$

we find that

$$\langle a^\dagger a\rangle_1 = \frac{1}{4}(2 + 4r^2 - 2)e^{r^2}|_r = 0 \tag{10.30}$$

Not too bad. How about for $|l = \rangle$?

We use the previous result for

$$\frac{\partial^2}{\partial^2 r}e^{r^2} \tag{10.31}$$

and find

$$\frac{\partial^4}{\partial^4 r}e^{r^2} = (8 + 40r^2 + 16r^4)e^{r^2} \tag{10.32}$$

to yield

$$\langle a^\dagger a\rangle_1 = \frac{1}{4}(8 + 40r^2 + 16r^4 - 4 - rr^2)e^{r^2}|_{r=0} = 1 \tag{10.33}$$

Due to the singular nature of the P distribution for interesting states, we will not be presenting any. Most of the work with the P distribution has been done with a linearization of the dynamics, where small nonclassicalities 'ride' along on a Gaussian distribution.

10.2 Wigner distribution

Another quasiprobability of utility is one used for calculating symmetrically ordered operators. We start with the characteristic function [9]

$$\chi_S = \mathrm{tr}(\rho e^{i(\beta a^\dagger + i\beta^* a)}) \tag{10.34}$$

This will be useful for calculating expectation values of symmetrically ordered operators like

$$(a^\dagger a)_S = \frac{1}{2}(aa^\dagger + a^\dagger a) \tag{10.35}$$

$$(a^{\dagger 2} a)_S = \frac{1}{2}(a^{\dagger 2} a + 2a^\dagger a a^\dagger + a^{\dagger 2} a) \tag{10.36}$$

The Wigner quasiprobability distribution is the Fourier transform of χ_S

$$W(\alpha, \alpha^*) = \frac{1}{\pi}\int d^2 z \chi_S e^{-iz\alpha} e^{-iz^*\alpha^*} \tag{10.37}$$

We can calculate our expectation values via several equivalent ways

$$\langle (a^{\dagger p} a^q)\rangle_S = \mathrm{tr}(\rho a^{\dagger p} a^q \tag{10.38}$$

$$= \frac{\partial^{q+p}}{\partial^p(iz^*)\partial^p(iz)}\chi_S|_{z=z^*=0)} \tag{10.39}$$

$$= \int d^2\alpha W(\alpha, \alpha^*)\alpha^{*p}\alpha^q \tag{10.40}$$

The relation between the Wigner function and matrix elements of the density operator is

$$W(x_\alpha, y_\alpha) = \frac{1}{\pi}\int dy \langle X + y|\rho|X - y\rangle e^{-2iYy} \tag{10.41}$$

where X and Y are our quadrature operators, $x_\alpha = \Im\mathrm{m}(\alpha)$ and $y_\alpha = \Im(\alpha)$. It was originally written in terms of position and momentum operators

$$W(x, p) = \frac{1}{\pi\hbar}\int dy \langle x + y|\rho|x - y\rangle e^{-2ipy} \tag{10.42}$$

We also have the relation

$$\rho = \frac{1}{\pi}\int d^2\alpha W(\alpha)T(\alpha) \tag{10.43}$$

where the operator $T(\alpha)$ is given by

$$T(\alpha) = \frac{1}{\pi}\int d^2\beta D(\beta)e^{i(z*a^\dagger + za)} \tag{10.44}$$

where $D(\beta)$ is our old friend the displacement operator. The P distribution gives us a weighted sum of coherent state density matrices. The Wigner function seems more unwieldy than that, but it has one major utility

$$\int dp W(x, p) = P(x) \tag{10.45}$$

$$\int dx W(x, p) = P(p) \tag{10.46}$$

where we then have probability densities for x and y. We can also measure the Wigner function using balanced homodyne detection. For a CAT scan in medical situations, one measures the x-ray absorption along a set of radial directions. From knowing the total absorption in all directions, one can reconstruct a 2D image of x-ray absorption via a radon transform. The same method applies for the Wigner function roughly. One measures the balanced homodyne signal for all choices of local oscillator phase. That set of signals is then radon transformed to yield the Wigner distribution. Just as in medical applications there are many new improvements beyond this technique for image reconstruction, but the basic method is the same. This was first demonstrated by Smithey *et al.*

What do Wigner functions look like? Let us start with Fock states. But first we show some relations between the Wigner function and the P distribution

$$W(\alpha) = \frac{2}{\pi}\int d^2\beta P(\beta)e^{-2|\alpha-\beta|^2} \tag{10.47}$$

$W(\alpha)$ is a convolution of the P distribution with a Gaussian, which acts as a smoothing filter.

Upon integration of the highly singular P_{Fock} with the Gaussian, and again using the identity for integration of a function against the derivative of a delta function, leading to evaluation of the derivative of the function at the location of the delta function, we find

$$W_n(\alpha) = \frac{2}{\pi}(-1)^n L_n(4|\alpha|^2)e^{-2|\alpha|^2} \tag{10.48}$$

This is easy to see by examining the Rodrigues formula for the Laguerre polynomials, L_n

$$L_n(x) = \frac{e^x}{n!}\frac{d^n}{dx^n}(x^n e^x) \tag{10.49}$$

In figure 10.1, we show the Wigner distribution for the vacuum state, a rather harmless looking Gaussian. It is a coherent state with zero amplitude.

In figure 10.2, we plot the Wigner function for $|n = 2\rangle$. You can see that it takes negative values around $\alpha = \sqrt{2}$. This negativity is indicative of a nonclassical state.

Figure 10.3 shows nonclassicallity and characteristic ripples of the Wigner function of a Fock state, this being $|n = 10\rangle$.

For a thermal state, the P distribution is a Gaussian, and the convolution with another Gaussian to find the Wigner function of course yields another Gaussian

$$W_{th} = \frac{1}{\pi(\bar{n} + 1/2)} e^{-|\alpha|^2/(\bar{n}+1/2)} \tag{10.50}$$

An example is shown in figure 10.4.

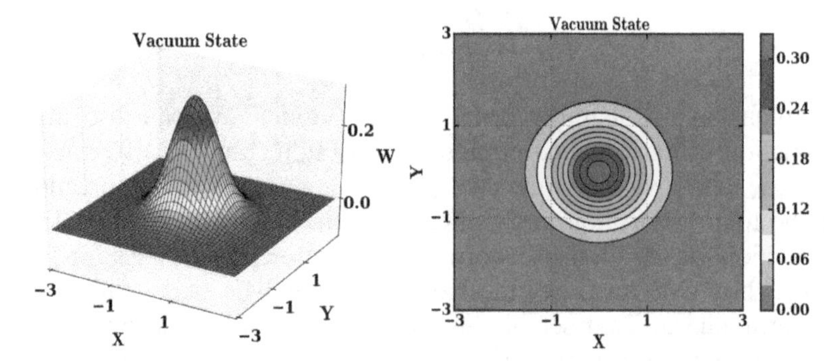

Figure 10.1. Wigner function for the vacuum state.

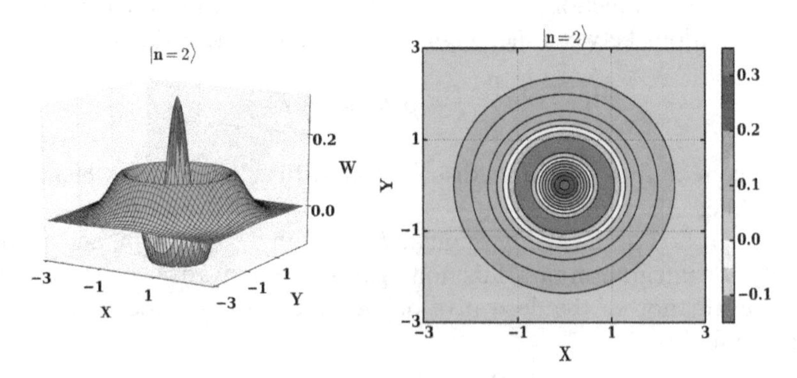

Figure 10.2. Wigner function for $|n = 2\rangle$.

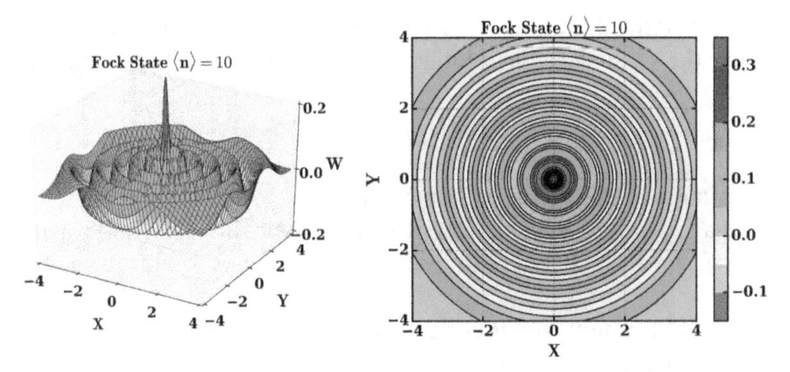

Figure 10.3. Wigner function for $|n = 10\rangle$.

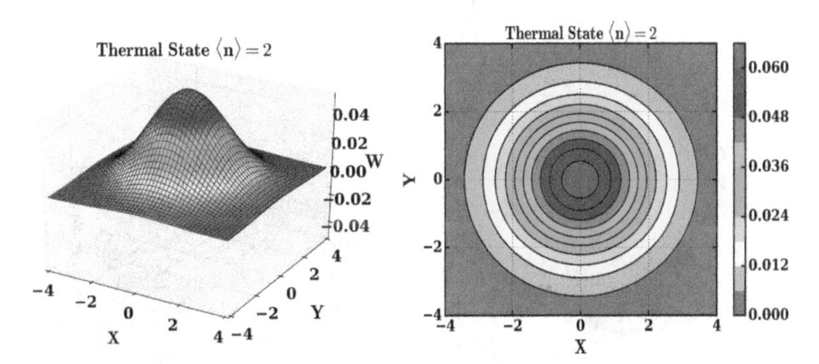

Figure 10.4. Wigner function for a thermal state with $\bar{n} = 10$.

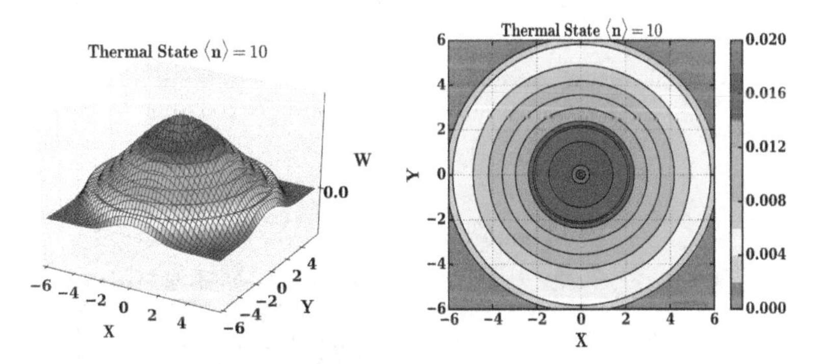

Figure 10.5. Wigner function for a coherent state with $|\alpha|^2 = 2$.

The Wigner distribution for a coherent state $|\alpha_0\rangle$ is the easiest to calculate as it is the convolution of a Gaussian with a δ function

$$W_{coh}(\alpha) = \frac{2}{\pi} e^{-|\alpha-\alpha_0|^2} \tag{10.51}$$

An example for $\alpha = 2$ is shown in figure 10.5.

To find the Wigner function for a squeezed vacuum state, it is again easier to obtain it from the Gaussian convolution with the corresponding P distribution

$$W_{sv}(x, y) = \frac{2}{\pi} e^{-2(e^{2r}x^2 - e^{-2r}p^2)} \tag{10.52}$$

where the Y quadrature is squeezed. An example for $|r = 10\langle$ is shown in figure 10.6. There are negative values that indicate it is nonclassicallity, beyond a width for the distribution of x values below 1/4. Squeezing in the X quadrature is shown in figure 10.7.

Increasing the amount of squeezing leads to more prominent ripples and negativity, as shown in figure 10.8.

Finally, we examine cat states of the form

$$|cat\rangle = \frac{1}{\sqrt{2}}(|\alpha\rangle + |-\alpha\rangle) \tag{10.53}$$

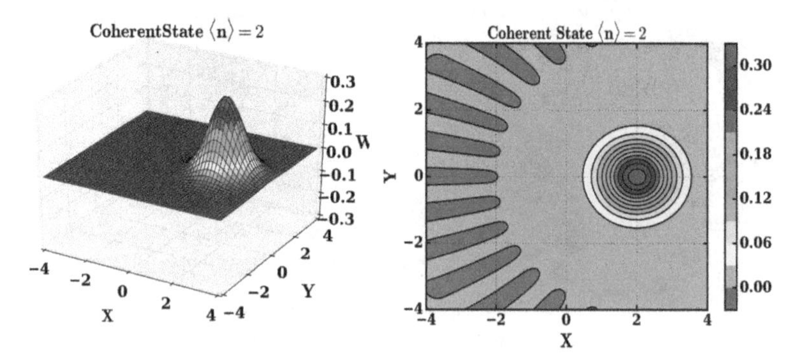

Figure 10.6. Wigner function for a squeezed vacuum state with $|\sinh(r)|^2 = 10$ with squeezing in the Y quadrature.

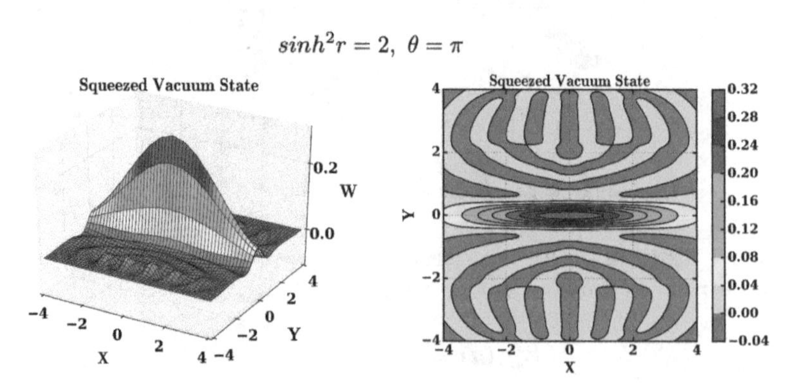

Figure 10.7. Wigner function for a squeezed vacuum state with $|\sinh(r)|^2 = 2$ with squeezing in the X quadrature.

The resulting Wigner function is a sum of two coherent state Wigner functions centered at α_0 and $-\alpha_0$, with an interference term given by

$$W_{int} = \frac{2}{\pi} e^{-2(e^{2r}x^2 - e^{-2r}p^2)} \cos(xp/4) \tag{10.54}$$

This interference term is negative sometimes and indicates that this sum of two essentially classical states is indeed nonclassical. This is shown for $\alpha = 2$ in figure 10.9, and for $\alpha = 10$ in figure 10.10.

Two other examples with the anti-cat state

$$|acat\rangle = \frac{1}{\sqrt{2}}(|\alpha\rangle|-\alpha\rangle) \tag{10.55}$$

are shown in figures 10.11 and 10.12.

10.3 Husimi Q function

The Husimi, or Q, function [3] is useful for calculating expectation values of anti-normally ordered operators such as

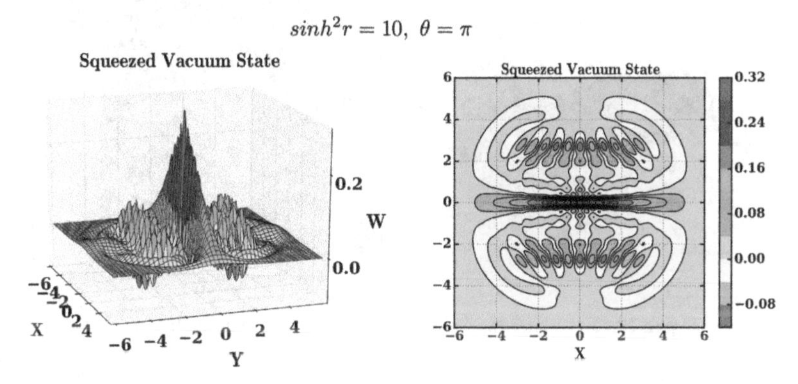

Figure 10.8. Wigner function for a squeezed vacuum state with $|\sinh(r)|^2 = 10$ with squeezing in the X quadrature.

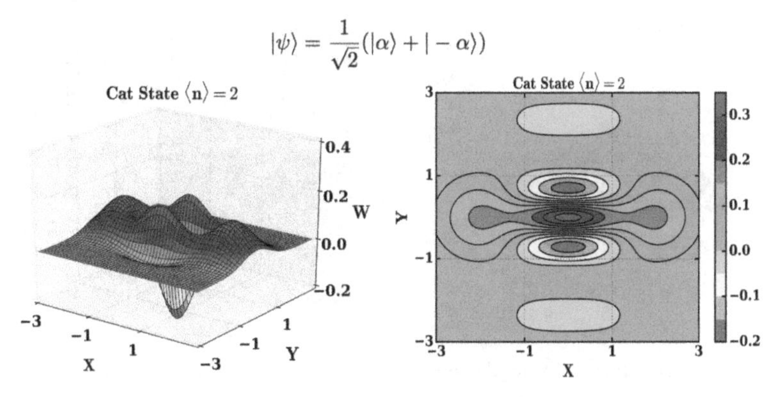

Figure 10.9. Wigner distribution for the cat state with $\alpha = 2$.

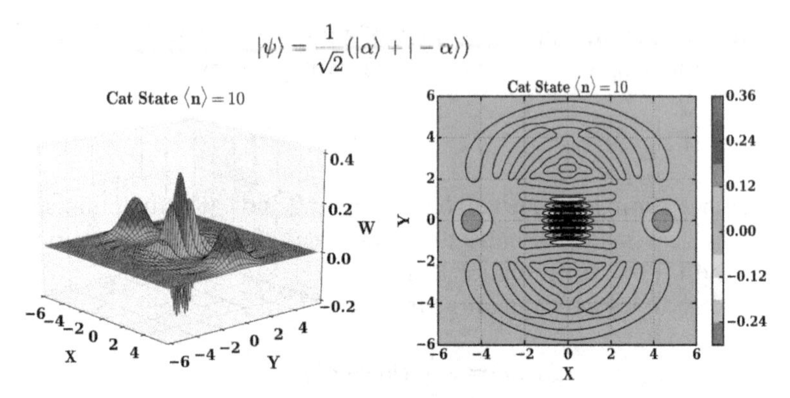

Figure 10.10. Wigner distribution for the cat state with $\alpha = 10$.

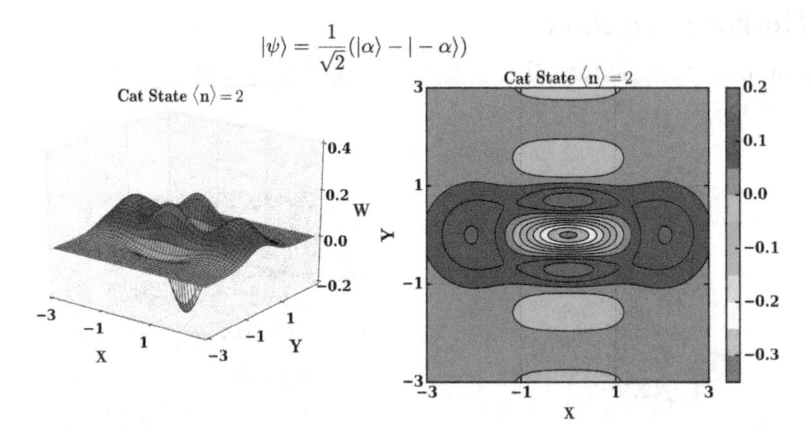

Figure 10.11. Wigner distribution for the anti-cat state with $\alpha = 2$.

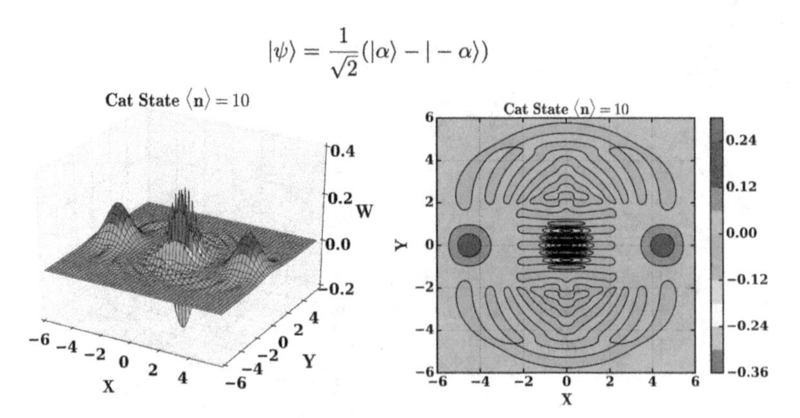

Figure 10.12. Wigner distribution for the anti-cat state with $\alpha = 10$.

$$(a^\dagger a)_A = aa^\dagger \tag{10.56}$$

The characteristic function for this distribution is

$$\chi_A = \mathrm{tr}(\rho e^{i\alpha^* a} e^{i\alpha a^\dagger}) \tag{10.57}$$

where we find

$$\langle a^n a^{\dagger m} \rangle = \left. \frac{\partial^{m+n} \chi_A}{\partial^n (i\beta) \partial^m (i\beta^*)} \right|_{\beta=\beta^*=0} \tag{10.58}$$

The Q function is given by diagonal matrix elements of the density matrix in the coherent state basis

$$Q(\alpha) = \langle \alpha | \rho | \alpha \rangle \tag{10.59}$$

As such, the Q distribution is positive definite and can be regarded as a true probability distribution.

We can calculate antinormally ordered expectation values via

$$\langle a^n a^{\dagger m} \rangle = \int Q(\alpha) \alpha^{*m} \alpha^n d^2\alpha \tag{10.60}$$

The Q distribution can also be written as convolutions of Gaussians with P or W distributions

$$Q(\alpha) = \frac{1}{\pi} \int d^2\beta P(\beta) e^{-|\alpha-\beta|^2} \tag{10.61}$$

$$= \frac{2}{\pi} \int d^2\beta W(\beta) e^{-2|\alpha-\beta|^2} \tag{10.62}$$

Armed with these, we now look at some examples. For a Fock state, we find

$$Q_n(\alpha) = \frac{-|\alpha|^{2n}}{\pi n!} e^{-|\alpha|^2} \tag{10.63}$$

We see something interesting for the vacuum state, the Q function has support at $|\alpha| = 1$ as shown in figure 10.13. You would think for the vacuum the distribution would be peaked at $\alpha = 0$. However, recall it is used for expectation of anti-normally ordered operators, so

$$\langle (a^\dagger a)_A \rangle = \langle aa^\dagger \rangle = 1 \tag{10.64}$$

We see that $Q_1(\alpha)$ has similar support at $\alpha = 2$ as shown in figure 10.14.

For a thermal state, we find

$$Q_{th} = \frac{1}{\pi(\bar{n} + 1)} e^{-|\alpha|^2/(\bar{n}+1)} \tag{10.65}$$

We plot an example with $\langle n \rangle = 2$ in figure 10.15.

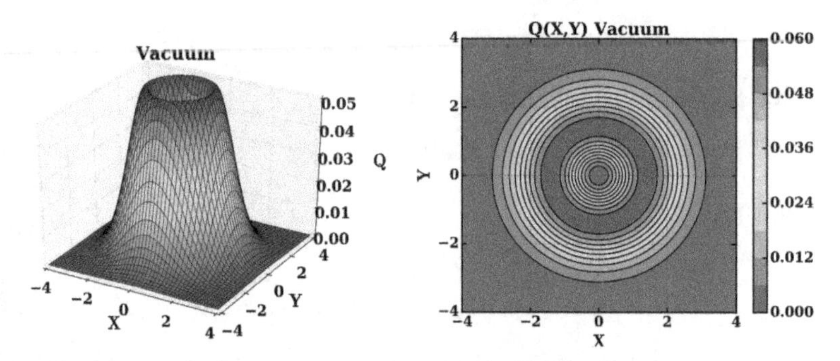

Figure 10.13. Q function for the vacuum state.

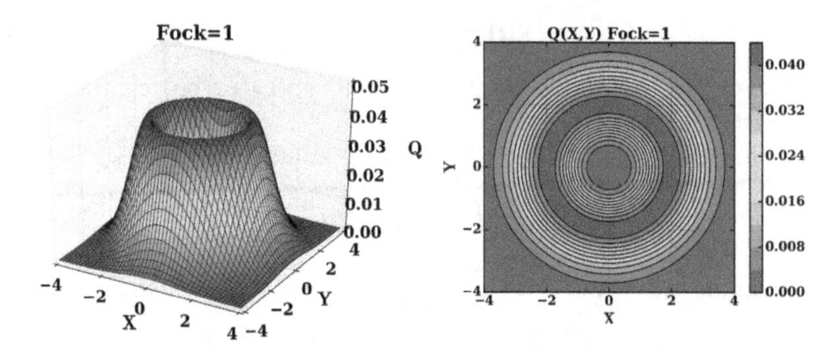

Figure 10.14. Q function for the single photon state.

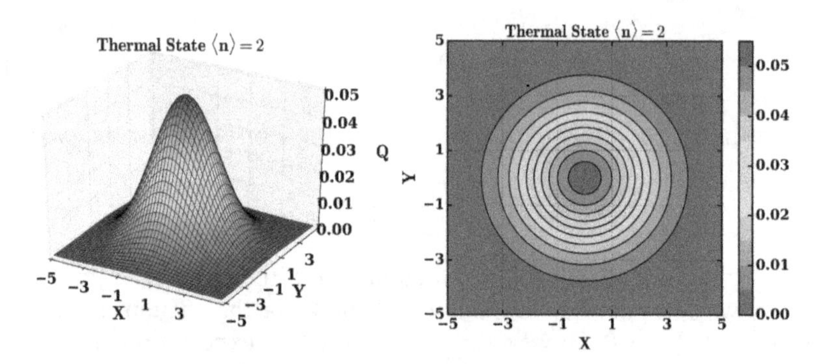

Figure 10.15. Q function for a thermal state with $\langle n \rangle = 2$.

For a coherent state, we merely obtain a Gaussian of the form

$$Q_{coh} = e^{-|\alpha|^2} \tag{10.66}$$

which we exhibit for a coherent state with $\langle n \rangle = 2$ in figure 10.16. It is rather indistinguishable from a thermal state.

For a squeezed vacuum state, we find no pleasant closed form. We show the Q function for squeezed vacuum states in figure 10.17, and in figure 10.18 for squeezing

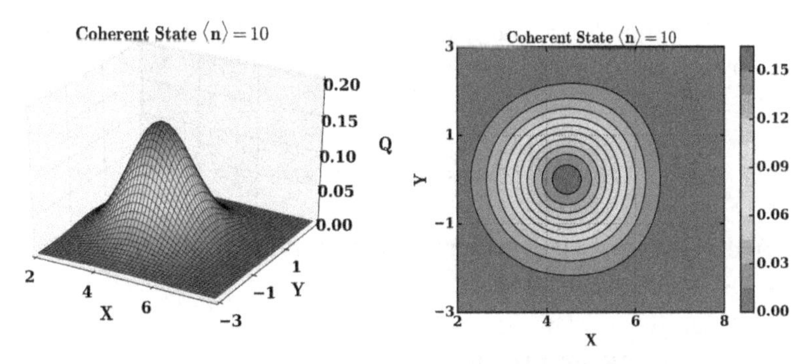

Figure 10.16. Q function for a coherent state with $\langle n \rangle = 2$.

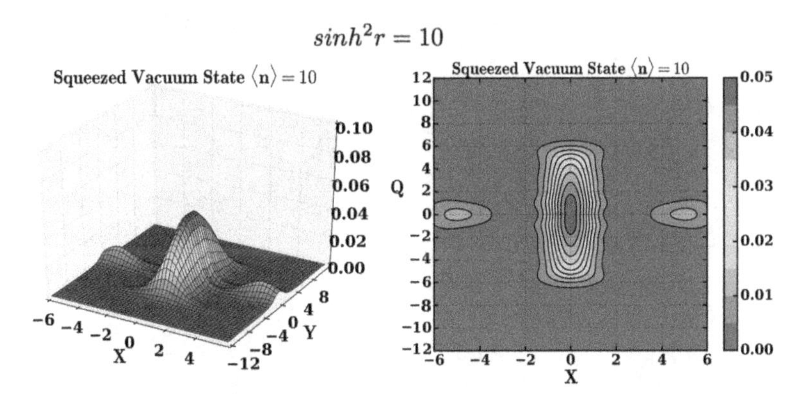

Figure 10.17. Q function for a squeezed vacuum state with $\langle n \rangle = 2$.

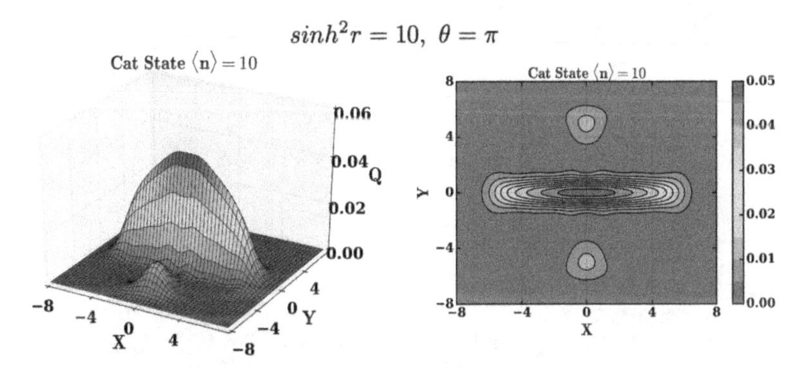

Figure 10.18. Q function for a squeezed vacuum state with $\langle n \rangle = 2$ with squeezing in the other quadrature.

in the other quadrature. One can indeed notice the decrease in fluctuations in one quadrature at the expense of the other.

For our last exploration of the Q function we examine cat states of the form

$$|cat\rangle = \frac{1}{\sqrt{2}}(|\alpha\rangle + |-\alpha\rangle) \tag{10.67}$$

We see essentially a double-peaked distribution in figure 10.19 that has constructive interference in the center, this is where the nonclassicallity is hidden.

For the anti-cat state

$$|acat\rangle = \frac{1}{\sqrt{2}}(|\alpha\rangle + |-\alpha\rangle) \tag{10.68}$$

we again see a two-peaked structure but this time with negative interference apparent in the center (figure 10.20).

10.4 Fokker–Planck equations

What about the time evolution of a quasiprobability distribution? We know we have the master equation for the density operator, say for a field mode coupled to a thermal reservoir

$$\begin{aligned}\dot{\rho}(t) = &-i\omega_0[a^\dagger a, \rho] + \kappa(\bar{n} + 1)[2a\rho(t)a^\dagger - \rho(t)a^\dagger a - a^\dagger a\rho(t)] \\ &+ \kappa\bar{n}[2a\rho(t)a^\dagger - aa^\dagger\rho(t) - \rho(t)aa^\dagger],\end{aligned} \tag{10.69}$$

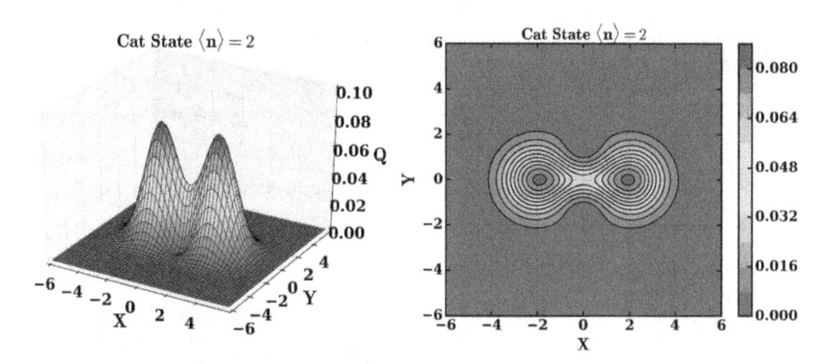

Figure 10.19. Q distribution for the cat state with $\alpha = 2$.

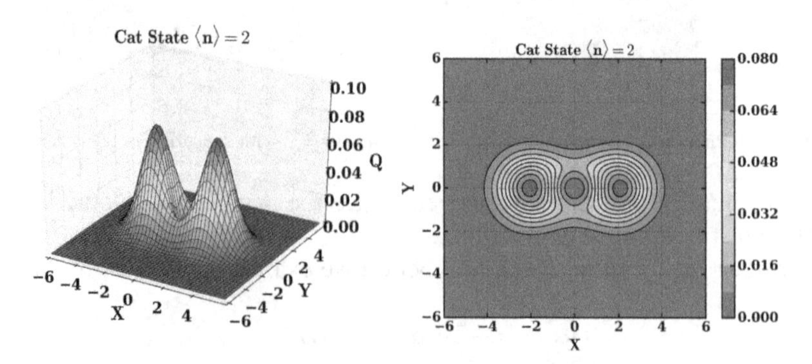

Figure 10.20. Q distribution for the anti-cat state with $\alpha = 2$.

If we replace ρ by its P representation, we see how the left side of the master equation yields

$$\dot{\rho} = \int d^2\alpha \dot{P}|\alpha\rangle\langle\alpha| \tag{10.70}$$

What happens to the right-hand side? Well we know if we have terms with $a|\alpha\rangle = \alpha|\alpha\rangle$ or the conjugate they are easy to deal with. But what about $a^\dagger|\alpha\rangle$?

$$a^\dagger|\alpha\rangle = a^\dagger e^{-\alpha\alpha^*} e^{\alpha a^\dagger}|0\rangle\langle 0|e^{\alpha^* a} \tag{10.71}$$

We can produce an $a^\dagger$ in front of $|0\rangle$ via a derivative with respect to α, and in fact

$$\frac{\partial}{\partial\alpha} e^{-\alpha\alpha^*} e^{\alpha a^\dagger}|0\rangle\langle 0|e^{\alpha^* a} = (a^\dagger - \alpha^*) e^{-\alpha\alpha^*} e^{\alpha a^\dagger}|0\rangle\langle 0|e^{\alpha^* a} \tag{10.72}$$

So, we then have the following correspondences:

$$a\rho \rightarrow \alpha P(\alpha) \tag{10.73}$$

$$\rho a^\dagger \rightarrow \alpha^* P(\alpha) \tag{10.74}$$

$$\rho a \rightarrow \left(\alpha - \frac{\partial}{\partial\alpha^*} P(\alpha)\right) \tag{10.75}$$

$$a^\dagger\rho \rightarrow \left(\alpha^* - \frac{\partial}{\partial\alpha} P(\alpha)\right) \tag{10.76}$$

Plugging these into the master equation, we find

$$\frac{\partial P}{\partial t} = (\gamma + i\omega_0)\frac{\partial}{\partial\alpha}\alpha P + (\gamma - i\omega_0)\frac{\partial}{\partial\alpha^*}\alpha^* P \tag{10.77}$$

$$+ \frac{\gamma}{2}\frac{\partial^2}{\partial\alpha^*\partial\alpha^*} P \tag{10.78}$$

We can perform the same procedure with the W and Q distributions, using

$$a\rho \rightarrow \left(\alpha + \frac{1}{2}\frac{\partial}{\partial\alpha^*}\right) W(\alpha) \tag{10.79}$$

$$\rho a^\dagger \rightarrow \left(\alpha^* + \frac{1}{2}\frac{\partial}{\partial\alpha}\right) W(\alpha) \tag{10.80}$$

$$\rho a \rightarrow \left(\alpha - \frac{1}{2}\frac{\partial}{\partial\alpha^*}\right) W(\alpha) \tag{10.81}$$

$$a^\dagger\rho \rightarrow \left(\alpha^* - \frac{1}{2}\frac{\partial}{\partial\alpha}\right) W(\alpha) \tag{10.82}$$

and

$$a^{\dagger}\rho \rightarrow \alpha^{*}Q(\alpha) \tag{10.83}$$

$$\rho a \rightarrow \alpha Q(\alpha) \tag{10.84}$$

$$a\rho \rightarrow \left(\alpha + \frac{\partial}{\partial\alpha^{*}}Q(\alpha)\right) \tag{10.85}$$

$$\rho a^{\dagger} \rightarrow \left(\alpha^{*} + \frac{\partial}{\partial\alpha}Q(\alpha)\right) \tag{10.86}$$

The resulting equations for Q and W time evolution are the same as for P with the replacements

$$\langle n\rangle \rightarrow \langle n\rangle + 1/2 \tag{10.87}$$

for W, and

$$\langle n\rangle \rightarrow \langle n\rangle + 1 \tag{10.88}$$

for Q. The solution for these with an initial coherent state $|\alpha_0\rangle$ is given by

$$P/W/Q = \frac{1}{\pi}\frac{1}{K + \langle n\rangle(1 - e^{-2\gamma t})} \tag{10.89}$$

$$x \exp\left(-\frac{|\alpha - \alpha_0(\exp - (\gamma + i\omega_0)t\,|^2)}{K + \langle n\rangle(1 - e^{-2\gamma t})}\right) \tag{10.90}$$

with $K = 0$ for P, 1/2 for W, and 1 for Q.

The time evolution of a quasiprobability can often take the following form

$$\frac{\partial P}{\partial t} = -\left(\frac{\partial}{\partial x}A(x)P(x) + \frac{1}{2} - \frac{\partial^2}{\partial^2 x}D(x)P(x)\right) \tag{10.91}$$

which is of the form of a Fokker–Planck equation used in statistical mechanics. See the excellent review by Risken [6]. So, it looks classical except for the quantum weirdness hidden in the quasiprobability distribution itself. Often to obtain an equation in this form, it is necessary to make a small noise approximation, where we have a classical distribution with correction terms of the order of an inverse system size, say atom number or saturation photon number, to some power. We may solve for the expectation value and variance of x,

$$\langle\dot{x}\rangle = \langle A(x)\rangle \tag{10.92}$$

$$\langle\dot{x}^2\rangle = 2\langle A(x)x\rangle + \langle D(x)\rangle \tag{10.93}$$

leading to

$$\Delta\dot{x}^2 = 2\langle A(x)x\rangle - s\langle x\rangle\langle A(x)\rangle + \langle D(x)\rangle \tag{10.94}$$

In the case of a linear Fokker–Planck equation,

$$A(x) = Ax \tag{10.95}$$

$$D(x) = D \tag{10.96}$$

the solution is

$$P(x,\, t) = \frac{1}{4\pi Dt}\exp\left(-\frac{(x - At)^2}{4Dt}\right) \tag{10.97}$$

Here, we see that the resulting distribution is a Gaussian where the mean is centered at $x = At$, with a width $2Dt$ as in figure 10.21.

So, the mean moves at a drift rate A with a diffusion of the width of $2Dt$. Hence, A (or set of A's in multiple variables) are known as drift coefficients, and the set of D_{ij} are known as diffusion coefficients.

10.5 Fermions

So far, we have only discussed quasiprobability distributions of bosons, what about fermions? Well we can certainly construct a characteristic function

$$\chi_N = \mathrm{tr}(\rho e^{i\mu^*\sigma_+}e^{iz\sigma_z}e^{i\mu\sigma_-}) \tag{10.98}$$

Here, the N refers to normal ordering, lowering operators to the right and raising operators to the left just as for bosons. We can calculate expectation values via

$$\langle\sigma_+^p\sigma_z^q\sigma_-^r\rangle = \mathrm{tr}(\rho\sigma_+^p\sigma_z^q\sigma_-^r) \tag{10.99}$$

$$= \frac{\partial^{p+q+r}}{\partial^p(i\mu^*)\partial^q(iz)\partial^r(i\mu)}\chi_N|(\mu - \mu^* = z = 0) \tag{10.100}$$

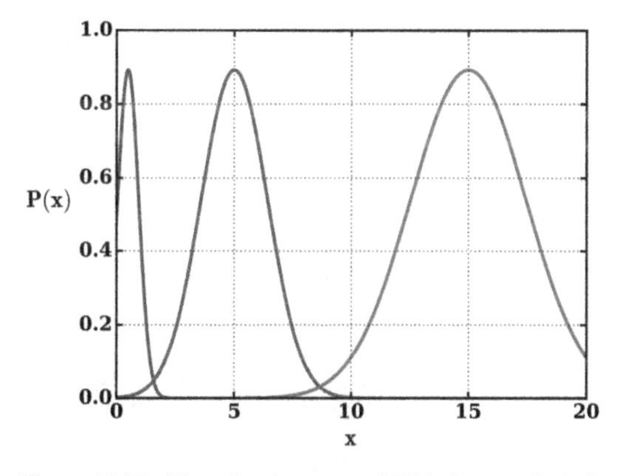

Figure 10.21. Time development of $P(x)$ for $t = .2,\, 1,\, 5$.

and write down a quasiprobability function

$$P(\mu, \mu^*, z) = \frac{1}{2\pi^3} \int d^2w d\eta \chi_N(w, w^*, \eta) e^{-iw\mu - w^*\mu^* - \eta z} \tag{10.101}$$

The difficulty arises when we convert this into a Fokker–Planck equation, that we have terms in the diffusion part of the form

$$e^{\partial/\partial z} \tag{10.102}$$

which having an infinite number of derivatives, which brings on enormous difficulties.

We can find atomic coherent states that are useful, however. They are states of minimum and symmetric noise. Consider the $|j, -j\rangle$ state of a spin. We know that $J_z = -j$ and that $J^2 = j(j+1)$, suppressing the $\hbar$s.

So, this is a state where

$$\Delta J_z = 0 \tag{10.103}$$

$$\Delta J_x \Delta J_y = \frac{j}{2} \tag{10.104}$$

with $\Delta J_x = \Delta J_y$. The atomic coherent states are just this state with symmetric, and minimum noise rotated to some other location on the Bloch sphere. This is shown schematically in figure 10.22.

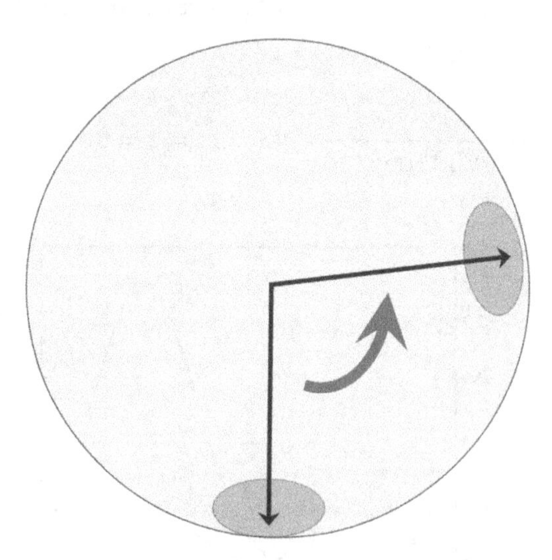

Figure 10.22. Schematic of atomic coherent states.

The phase uncertainty of these states is given by

$$\Delta\phi = \Delta J \langle J_z \rangle \approx 1/\sqrt{N} \tag{10.105}$$

where N is the number of atoms.

We can formalize this by defining [1]

$$|ACS\rangle = |\theta, \phi\rangle = \cos(\theta/2)|-\rangle + e^{i\phi}\sin(\theta/2)|+\rangle \tag{10.106}$$

in the single atom case. As with bosonic coherent states, we can generate these states with a type of displacement operator, or rotation operator

$$|\theta, \phi\rangle = \exp(i(zJ_+ + z^*J_-))|j, -j\rangle \equiv R(z)|j, -j\rangle \tag{10.107}$$

where $|j, -j\rangle$ is the state with

$$J_z|j, -j\rangle = -j|j, -j\rangle \tag{10.108}$$

$$J^2|j, -j\rangle = j(j+1)|j, -j\rangle, \tag{10.109}$$

the state of N atoms with the lowest z component of angular momentum. The rotation parameter z is given by

$$z = -\frac{\theta}{2}e^{-i\phi} \tag{10.110}$$

Atomic squeezed states can be defined as above, with minimum noise $\Delta J_x \Delta J_y = j/2$ but with

$$\Delta J_x = e^{-2r}\sqrt{j/2} \tag{10.111}$$

$$\Delta J_y = e^{2r}\sqrt{j/2} \tag{10.112}$$

A more formal definition of a squeezing parameter, by analogy with bosonic squeezed states, is [5]

$$s = 2(\Delta J_i)^2/\langle J_k \rangle \tag{10.113}$$

with $j \neq k$.

Unfortunately, this parameter can be less than unity, implying squeezing, for an atomic coherent state. The analogy with bosonic squeezed states breaks down due to a preferred direction, given by $\langle \vec{J} \rangle$. This stems from the fact that these states live on a Bloch sphere instead of a plane. Wineland has defined a useful squeezing parameter for metrology as [10]

$$s_W = N(\Delta J_{\vec{n}\perp})^2/\langle \vec{J} \rangle^2 \tag{10.114}$$

where $\vec{n}_\perp \cdot \langle \vec{J} \rangle = 0$. This always yields the ration

$$s_W = \frac{\Delta\phi^2}{\Delta\phi_{ACS}^2} \tag{10.115}$$

Spin squeezing is shown schematically in figure 10.23.

Due to the interactions of spin with magnetic fields, these latter states can be utilized for quantum enhanced magnetic field measurements.

We can make an analogy with bosonic squeezed states, beginning with the Jordan–Schwinger representation

$$\vec{J} = A^{\dagger}\vec{\sigma}A \tag{10.116}$$

with

$$A = \begin{pmatrix} a \\ b \end{pmatrix} \tag{10.117}$$

with a and b bosonic operators satisfying the usual bosonic commutation operators, states of definite angular momentum are given by

$$|j, m\rangle = \frac{a^{\dagger k} b^{\dagger n}}{k!n!}|0\rangle \tag{10.118}$$

with $j = (k + n)/2$ and $m = (k - n)/2$. We have the transformation

$$J_x = (a^{\dagger}b + b^{\dagger}a)/2 \tag{10.119}$$

$$J_y = (a^{\dagger}b - b^{\dagger}a)/2 \tag{10.120}$$

$$J_z = (a^{\dagger}a + b^{\dagger}b)/2 \tag{10.121}$$

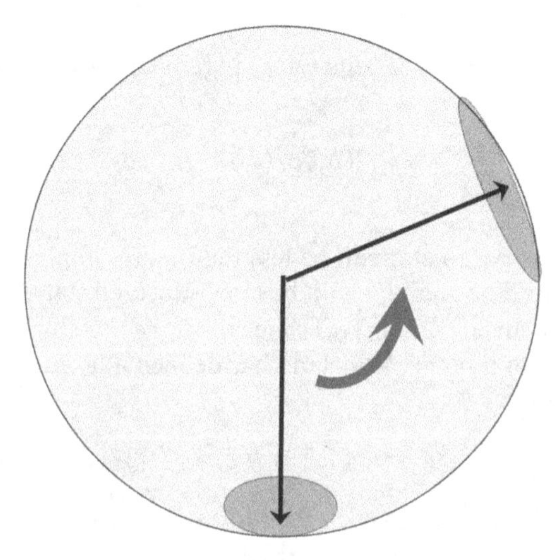

Figure 10.23. Schematic of spin squeezing.

An interaction of the form J_x^2 or J_y^2 will contain terms of the form

$$a^{\dagger 2}b^2 + \text{h.c.} \tag{10.122}$$

which are analogous to the bosonic squeezing operators. These are known as twisting interactions.

10.6 Langevin equations

For our purposes, a Langevin equation is a Heisenberg equation of motion for a set of operators, with a noise term representing interactions with a bath, as well as a decay/gain term arising from those same interactions. More properly called Heisenberg–Langevin. As a simple example, let us consider a single field mode interacting with a continuum of field modes

$$H = \hbar\omega_0 a^\dagger a + \sum_k \hbar\omega_k b_k^\dagger b_k + \sum_k g_k(ab_k^\dagger + a^\dagger b_k) \tag{10.123}$$

The Heisenber equations of motion are given by

$$\dot{a} = -i\omega_0 a - i\sum_k g_k b_k \tag{10.124}$$

$$\dot{b}_k = -i\omega_k b_k - ig_k a \tag{10.125}$$

Substituting the equations for $\dot{b}_k$ into the equation for $\dot{a}$, we have

$$\dot{a} = -i\omega_0 a - i\sum_k |g_k|^2 a(t')e^{-i\omega_k(t-t')} + f_a(t) \tag{10.126}$$

where we have defined the noise operator

$$f_a(t) = -i\sum_k g_k b^k(0)e^{-i\omega_k t} \tag{10.127}$$

Here, we have replaced $b_k(t)$ by $b^k(0)e^{-i\omega_k t}$, keeping only lowest order in the (assumed small) coupling constant g_k. Similarly, we can replace

$$a(t') \to a(t)e^{-i\omega_0(t'-t)} \tag{10.128}$$

in the field mode equation. This is exactly what we did back in chapter 6 when we discussed vacuum fluctuations and radiation reaction, as well as deriving the Wigner–Weiskopff formula. This results in

$$\dot{a} = -i\omega_0 a - \Gamma(t')e^{-i\omega_k(t-t')} + F_a(t) \tag{10.129}$$

where we have

$$\Gamma = 2\pi\sum_k |g_k|^2 \tag{10.130}$$

and

$$F_a = -i\sum_k b_k(0)e^{-i(\omega_k-\omega_0)t} \tag{10.131}$$

If one ignores the fluctuating noise operator, F_a, the solution is given by

$$a(t) = a(0)e^{-\Gamma t/2} \tag{10.132}$$

The difficulty is that the commutator between a and $a^\dagger$ will decay to zero! To have decay/gain, one must have fluctuations. This is the heart of the celebrated fluctuation–dissipation theorem.

The noise term has zero mean, but a nonzero intensity

$$\langle F_a \rangle = 0 \tag{10.133}$$

$$\langle F_a^\dagger(t)F_a(t')\rangle = \Gamma\delta(t - t') \tag{10.134}$$

As the noise is completely different from one time to another, even with very small separation, unless $t = t'$, the correlation is zero. If $t = t'$, you are correlating something with itself, and over time that just grows, hence the δ function. To simulate this behavior, we write

$$da = -(i\omega_0 + \Gamma/2)da + \sqrt{\Gamma}dW \tag{10.135}$$

Here, dW is what is known as a Weiner process, and it is simulated by drawing a random number from a Gaussian distribution of unit mean and zero average, and multiplying by the square root of the time step.

$$dW_{sim} = r\sqrt{dt} \tag{10.136}$$

where r is the random number drawn from the Gaussian distribution. The reason for the square root is that this is essentially a diffusive process, where the intensity of the noise grows linearly with time. It is an odd process. Each time step it jumps randomly, and as such it has no meaningful derivative. This is why we never write dW/dt. A simulation for a nonzero initial amplitude will decay noisily to zero, an average of such simulations will yield the result for the expectation value, as is shown in figure 10.24.

$$\langle a(t)\rangle = \langle a(0)\rangle e^{-\Gamma t/2} \tag{10.137}$$

Let us now examine the Weiner process in more detail [4], following the elegant treatment of Steck [7]. If we have a sum of random numbers X with average $\langle X\rangle = 0$ and variance $\Delta X \equiv \langle X^2\rangle - \langle X\rangle^2 = \sigma$, what is the behavior of that sum? This is essentially the random walker who takes random steps left or right. You might expect that after N steps the walker is at $\langle X\rangle = 0$ but they are not. The central limit theorem says that the probability distribution for the position after N steps is given by a Gaussian with zero mean and variance $\sqrt{N}\sigma$. If we have N steps of Δt, with $N\Delta t = t$, we have a standard deviation, the square root of the variance, given by

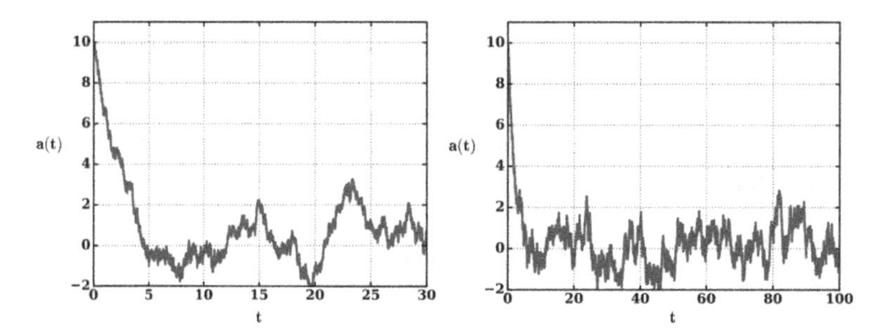

Figure 10.24. Solution of a damped harmonic oscillator for short and long times.

$$\sigma(t) = \sigma\sqrt{t/\Delta t} \tag{10.138}$$

This is a diffusive process with a root mean square displacement of

$$\Delta x(t) = D\sqrt{t} \tag{10.139}$$

In this example, the diffusion coefficient depends on the time step, $D = \sigma/\sqrt{\Delta t}$.

A key fact is that for random steps, even if they are not identically distributed, it is the variance that adds, and not the average. If $X = X_1 + X_2$, then

$$(\Delta X)^2 = (\Delta X_1)^2 + (\Delta X_2)^2 \tag{10.140}$$

If the distributions are identical, with variance σ, we obtain once again

$$\sigma_N = N\sigma_0 \tag{10.141}$$

What if our steps are continuous and not discrete? This is the Weiner process, $W(t)$, which satisfies a distribution

$$P(W(t)) = \frac{1}{2\pi t}e^{-W^2/2t} \tag{10.142}$$

With the choice $W(0) = 0$, we have

$$P(W(0)) = \delta(0) \tag{10.143}$$

We can then write the variance

$$\langle(\Delta W)^2\rangle = \Delta t \tag{10.144}$$

So, we can say that the Weiner increment

$$Q(t + dt) - W(t) = \sqrt{\Delta t} \tag{10.145}$$

This all suggests that in differential equations with random numbers, we must keep second order in dW to be consistent with keeping terms linear in dt. In particular

$$\Delta W/dt = 1/\sqrt{\Delta t} \tag{10.146}$$

which tends to infinity in the limit of $\Delta t \to 0$.

Consider now

$$dy = \alpha dt + \beta dW \tag{10.147}$$

a variable that changes in time and also has a random component. A function of y, say $z = e^y$, has derivative

$$\frac{d}{dt}z = e^{y(t+dt)} - y(t) \tag{10.148}$$

$$= z(dy + dy^2)/2 \tag{10.149}$$

where we keep the dy^2 due to the need to keep dW^2 to keep terms of order dt. The result is

$$dz = (\alpha + \beta^2/2)dt + z\beta dW \tag{10.150}$$

We can see how a decay/gain term γ, we must have noise on the order of $\sqrt{\gamma}$. If there is gain/dissipation, there will be noise. This form is known as the Ito form of a stochastic differential equation. There are other forms. In general, we have

$$y(t + dt) = y(t) + \alpha(Y(t), t)dt + \beta(Y(t + \tau, \tau))dW \tag{10.151}$$

with τ in the interval $t \to t + dt$. The Ito choice takes $\tau = t$. Another choice is the Stratonovich approach takes $\tau = t + dt/2$.

One major difference between the two approaches is what the chain rule for stochastic differentiation is. For the Ito calculus, we have

$$dz = df(y) = f'dy + (1/2)f''dy^2 \tag{10.152}$$

$$= [f'\alpha + (1/2)f''\beta]dt + f'\beta dW \tag{10.153}$$

which is not the usual chain rule. One of the appeals of the Stratonovich approach is that

$$df = f'\alpha dt + f'\beta dW \tag{10.154}$$

which is the usual chain rule. However, because things are evaluated at the middle of the interval between points, the strangeness is hidden. One must be careful to specify at the beginning whether one is using the Ito or Stratonovich approach and proceed consistently.

References

[1] Arecchi F T, Courtens E, Gilmore R and Thomas H 1972 Atomic coherent states in quantum optics *Phys. Rev. A* **6** 2211

[2] Glauber R J 1963 Coherent and incoherent states of the radiation field *Phys. Rev.* **131** 2766

[3] Husimi K 1940 Some formal properties of the density matrix *Proc. Phys. Math. Soc. Jpn* **22** 264

[4] Van Kampen N G 2007 *Stochastic processes in physics and chemistry* (Amsterdam: North-Holland)

[5] Ma J, Wang X, Sun C P and Nori F 2011 Quantum spin squeezing *Phys. Rep.* **509** 89
[6] Risken H 1996 *The Fokker-Planck equation: Methods of Solutions and Applications* Springer Series in Synergetics (Berlin: Springer)
[7] Steck D Atom and quantum optics notes (course notes) http://atomoptics-nas.uoregon.edu/~dsteck/teaching/quantum-optics/
[8] Sudarshan E C G 1983 Equivalence of semiclassical and quantum mechanical descriptions of statistical light beams *Phys. Rev. Lett.* **10** 277
[9] Wigner E P 1932 On the quantum correction for thermodynamic equilibrium *Phys. Rev.* **40** 749
[10] Wineland D J, Bollinger J J, Itano W M, Moore F L and Heinzen D J 1992 Spin squeezing and reduced quantum noise in spectroscopy *Phys. Rev. A* **46** R6797